Renewable Energy and its Innovative Technologies

Editors
Jayeeta Chattopadhyay
Department of Chemistry
Amity University
Ranchi, Jharkhand, India

Om Prakash
Department of Mechanical Engineering
National Institute of Technology Patna
Patna, Bihar, India

Rahul Singh
Department of Mechanical Engineering
Amity University
Ranchi, Jharkhand, India

ISBN 978-981-13-2115-3 ISBN 978-981-13-2116-0 (eBook)
https://doi.org/10.1007/978-981-13-2116-0

Library of Congress Control Number: 2018951199

This Springer imprint is published by the registered company Springer Nature Singapore Pte Ltd.
The registered company address is: 152 Beach Road, #21-01/04 Gateway East, Singapore 189721, Singapore

Preface

The book entitled *Renewable Energy and its Innovative Technologies* presents the innovative scientific research works evaluated by scientists, academicians, research scholars, and students and presented in the International Conference on Energy, Materials and Information Technology held at Amity University, Jharkhand, India, on December 23–24, 2017. Remove fossil fuels from the human equation and modern industrial civilization would cease to exist. The transformation to a fossil fuel civilization occurred more rapidly than any other changes in energy regimes in world history. Power generation through various renewable energy resources is the most promising challenge in the present day. From solar energy to wind energy and from biomass energy to hydro energy, all the areas are equally required to be explored with a more effective way, to make them the most reliable alternatives to the fossil fuel energy. This book aims at all those who are interested in physical and technical principles of promising ways to utilize various renewable energies. It will be designed to cover all the promising renewable energies and their related technologies, viz. wind energy, solar energy, biomass energy, to address a multidisciplinary scientific aspect of education in this subject. The main focus of this book will be to assemble the important scientific research works going on in this field and find ways for their application through interdisciplinary manner. All the contributing authors of this book have provided innovative and important data and parameter sets for the utilization of major possibilities of renewable energies which allow for an economic and environmental assessment. Such an assessment enables us to judge the chances and limits of the multiple options in utilizing renewable energy sources.

Ranchi, Jharkhand, India	Jayeeta Chattopadhyay
Ranchi, Jharkhand, India	Rahul Singh
Patna, Bihar, India	Om Prakash

Acknowledgements

It is our pleasure to present this volume consisting of selected papers based on the oral presentation in the International Conference on Energy, Materials and Information Technology at Amity University, Jharkhand, Ranchi, India, held on December 23–24, 2017. We would like to take this opportunity to thank all the participants of the conference—invited speakers, presenters, and audience alike. Our special thanks to all the keynote speakers present at the conference. We would like to thank SERB-DST for financial sponsorship for this conference. Heartfelt thanks to our Founder President Dr. Ahok K. Chauhan and Chancellor Dr. Atul Chauhan for showing us the right path. We would like to thank Prof. (Dr.) R. K. Jha, Vice Chancellor, Amity University, Jharkhand, for his enormous support. We thank Prof. (Dr.) Ajit Kr. Pandey, Director, Amity University, Jharkhand, for his valuable guidance. Here we would like to specially thank Dr. Nitya Garg and Mr. Pravin Singh without whom this proceeding would not have taken proper shape. We would like to thank all the faculty, staff, and students of Amity University, Jharkhand, for their continuous support in making this conference successful.

Contents

About the Editors

Jayeeta Chattopadhyay received her bachelor of science with honors in chemistry from Bethune College, Calcutta University. She got her master of science in chemistry from Devi Ahilya Vishwavidyalaya, Indore (2003), and master of technology in fuels and combustion from Birla Institute of Technology, Mesra, Ranchi (2005). She obtained her Ph.D. in new energy engineering with Best Doctoral Thesis Award from Seoul National University of Science and Technology, South Korea (2010). She has obtained the prestigious Fast Track Young Scientist Award (2010) and Early Career Research Award (2017) from the Department of Science and Technology, Government of India. She has published more than 20 research and review articles in high-impact, peer-reviewed, international journals, and she is the owner of one international patent. She is the reviewer for more than ten high-impact international journals and has been included in the editorial board of two international journals. Her research interest includes nano-structured materials for energy applications and thermo-degradation of solid waste materials. She also works on oscillatory chemical reactions and pattern formation in reaction–diffusion system. Earlier, she was working as Research Scientist in Birla Institute of Technology, Mesra, Deoghar Off-campus, Jharkhand. She is presently working as Senior Assistant Professor, Amity University, Jharkhand, Ranchi, India.

Mr. Rahul Singh graduated in mechanical engineering in 2009 from BIT; completed his postgraduate in machine design with distinction from BIT Mesra, Ranchi, in 2014; and is pursuing Ph.D. from National Institute of Technology, Patna, since 2014. He is now Assistant Professor in the Department of Mechanical and Automation Engineering at AMITY University, Jharkhand, Ranchi. He has served in various organizations like SSIPMT Raipur; Jain University, Belgaum; and Sikkim Manipal University, Gangtok. His specialized fields of research are design of machine components, heat and mass transfer analysis, renewable energy, assessment of wind power, and micro-wind turbine. He has done more than ten publications in the journals of national and international repute alongside conference proceedings. He has attended and organized seminars, workshops, and conferences of national and international repute. Apart from research publication, he has published four books on titles like *Fluid Mechanics, Fluid Machinery, and Dynamics of Machines*. His subjects of interest are fluid mechanics, heat and mass transfer, strength of materials, kinematics of machines, and machine design.

Prof. (Dr.) Om Prakash graduated in mechanical engineering in 1994, completed his postgraduate in design and production of thermal power equipment from NIT Trichy in 1997, and obtained his Ph.D. from Indian Institute of Technology Delhi in 2005. He is now Professor in the Department of Mechanical Engineering and Dean of Faculty Welfare at the National Institute of Technology, Patna. He has served in various organizations like Indian Air Force, Directorate of Training and Technical Education, Delhi College of Engineering, and Rajaram Sinde College of Engineering. His specialized fields of research are heat and mass transfer analysis, energy audit and energy efficiency, biogas-fueled diesel engine, and micro-wind turbine. He has done more than 30 publications in the journals of national and international repute. Apart from publication, he has guided more than 15 students for M.Tech. and 10 students (completed and ongoing) for Ph.D. He has attended and organized seminars, workshops, and conferences of national and international repute. He is a life member of professional bodies like SAEINDIA, Indian Society of Heating, Refrigerating and Air Conditioning Engineers (ISHRAE), Indian Society for Technical Education (ISTE), Institution of Engineers (India), and Society of Energy Engineers and Managers (SEEM).

Site-Specific Vulnerability Assessment of Buildings Exposed to Rockfalls

Aditi Singh, Shilpa Pal and D. P. Kanungo

Abstract A landslide is one of the most common hazards occurring in mountainous region in response to a number of natural and anthropogenic processes. In particular, South Asian nation including India is the worst hit by landslides. Thus the vulnerability assessment towards landslides becomes an important aspect, as it pose risk to human life, environment, buildings, and infrastructures. From the perspective of natural and engineering sciences, vulnerability is defined as the degree of loss, or damage to a set of element at risk within the affected area due to landslides. It acts as a primary component in the evaluation of landslide risk, and its accurate estimation is necessary in making a reasonable prediction of the landslide consequences. The present paper aims to produce a framework for assessing the physical vulnerability of building exposed to rockfall using the empirical formula proposed by Li et al. (Landslides 7(2):125–134, 2010 [3]) in terms of resistance offered by building and the intensity of rockfall. The proposed methodology permits to obtain an estimate of vulnerability of buildings when hit by three different intensities (low, medium and high) of rockfall. Finally, application of this proposed framework is illustrated through a case study from Chamoli district, Uttarakhand (India).

Keywords Landslide · Rockfall intensity · Resistance factor
Physical vulnerability

1 Introduction

Landslide are the most common natural hazards affecting large parts of India, especially the Himalayas, the Northeastern ranges, the Western Ghats, the Nilgiris, the

A. Singh (✉) · S. Pal
Civil Engineering Department, Gautam Buddha University, Greater Noida, India
e-mail: aditi.peace@gmail.com

D. P. Kanungo
Geotechnical Engineering Group, CSIR—Central Building Research Institute (CBRI), Roorkee, Uttarakhand, India

J. Chattopadhyay et al. (eds.), *Renewable Energy and its Innovative Technologies*,
https://doi.org/10.1007/978-981-13-2116-0_1

Eastern Ghats and the Vindhyas. The most vulnerable to landslides are the Himalayas and Western Ghats regions. Therefore, the built-up areas in these regions are also affected by landslides on a variety of spatial and temporal scales [1]. The damage to built environment such as buildings and other vital infrastructures (bridges, dams, etc.) due to the occurrence of rapid landslides (e.g., debris flow) and generally rockfalls can be most severe causing complete destruction of buildings within the affected area. Slow-moving slides may also lead to damaging impact in the affected area [2].

Therefore, it is necessary to consider a methodology that analyzes the magnitude of damage due to landslide occurrences in order to assess the consequences of landslide impact on humans and their activity since the methodologies to assess a building's vulnerability subjected to rockfalls are scarce. Therefore, a framework has been proposed for assessing the vulnerability of buildings exposed to rockfall based on the empirical formula proposed by Li et al. [3]. The present paper focuses on the previously developed methodology to assess physical vulnerability of buildings in the mountainous region using rockfall intensity and the resistance offered by building has been done. Li et al. [3] proposed a quantitative model for vulnerability of structures based on landslide intensity and the resistance of exposed elements, but the developed model was limited to only debris flow type of movement. To overcome this limitation, a new methodology has been proposed specifically for rockfall type, to study the impact of different intensities on the buildings. Further, the methodology has been illustrated by studying six buildings located on Chamoli district of Uttarakhand state.

2 Vulnerability of Buildings Due to Rockfall Impacts

A rockfall is generated by the detachment of a rock volume from the slope that falls, rolls, and bounces along its trajectory and may occur singly or in clusters. Rockfalls can be very destructive due to high velocities and there high energies that can reach during their downslope propagation [4]. Nevertheless this loss can vary for small to medium-sized events. Thus, the houses, roads, and bridges constructed in mountainous areas which are exposed to the danger of rockfalls may suffer from significant structural and non-structural damage. Although many methods exists for the analysis of the stability of rocky slopes [5], but the vulnerability assessment of buildings due to rockfalls is poor through empirical relations which results in uncertain loss predictions [6]. Empirical evaluations are mostly quantified without differentiating between typological classes for assessing rockfall vulnerability [1, 7]. Mavrouli and Corominas [8, 9] and Mavrouli et al. [10] presented a method for the assessment of vulnerability of reinforced concrete (RC) buildings. However, these results cannot be generalized and transferred to different structural typologies from the studied ones, as the structural performance of a building depends on various factors such as geometry, materials, and the load-bearing system of the elements.

Heinimann [11] and Uzielli et al. [12] presented a vulnerability value for buildings at regional scale according to the typology. Dai et al. [13] proposed a general framework for the classification of the physical vulnerability of buildings to landslides, at a global and site-specific scale, taking into account their individual characteristics (such as age, nature, and type). The methodology requires past event data for the statistical analysis. During the Fiumelatte event, Agliardi et al. [7] proposed a vulnerability function for rockfalls based on observations and back analysis of the damage of individual buildings.

Several rockfall events in Andorra have been reported by Corominas et al. [14, 15], during one such event of 1997, a block of 25 m^3 volume penetrated the slabs of a residence building, ending down to its basement. Further, the building did not suffer any extensive damage. Another such event occurred in 2004, Segovia, Spain, when boulders fell on the roof of the Fuencisla Sanctuary [16]. Additionally various events of landslide have been registered at Hong Kong [17], Canada [18] and United States [19, 20]. Since, empirical or heuristic approaches for the quantification of the vulnerability, requires historical recorded data for rockfalls, which are rarely available. Implementation of numerical and analytical models can be an appropriate alternative for the quantification of vulnerability.

3 Description of Method

According to UNDP/UNESCO 1982 [21], four main groups of methods exist for assessing the level of vulnerability of a structure: (i) categorization methods, which is based on the typological classification of structures into typological classes; (ii) inspection and empirical (rating) methods, where numerical value is attributed to each structure; (iii) analytical methods, based on the analysis of a structure depending on the resistance offered at an hazard event; and (iv) experimental methods, includes tests for the determining the structural properties of the whole structure and its components. In this study, a framework has been developed taking into account method (i), (ii) and (iii). These three methods are integrated together to assess site-specific vulnerability of buildings exposed to rockfall.

The very first step is to categorize different typologies of building. Since, every country follows their unique structural typology, construction material, and construction techniques depending on the geomorphological and geological conditions of that particular area. In India, the most common type of structural typology in Himalayan region is reinforced concrete, brick masonry and stone masonry. Second step involves the inspection of the structure and then assigning numerical value to each structure using rating (empirical) methods, therefore the buildings are first inspected and then numerical value is assigned based on the rating scheme developed. In the third step, resistance of buildings is calculated using an empirical formula. Finally, by integrating all three steps vulnerability of the buildings is calculated.

poor workmanship. Therefore it is an important parameter for assessing physical vulnerability [1].

(iii) *State of maintenance*: It is another important parameter for all the structural and non-structural components. Since it provides resistance to buildings in reducing the impact of landslides. In this study, state of maintenance is expressed in five classes as: very poor, poor, medium, good, very good.

(iv) *Number of floors*: The number of floors (or storeys) in a building also plays a crucial role in deciding the response of buildings to landslides. In the Indian Himalayas, the maximum three floors of buildings are observed. Hence, in this study three different floors were distinguished as Ist Floor, IInd Floor, and IIIrd Floor. Here, in case of rockfall we assumed that the top-most floor of the building will suffer more damage due to the impact of unstable fragments of rocks than those having two or more floors. However, in case debris flow which can be rapid landslides, taller buildings having three or more floors are likely to suffer less damage than low-rise buildings (i.e., one floor). This is due to the fact that intrusion of material in low-rise building is easy through the openings than in the high-rise buildings.

Thus resistance of the buildings exposed to landslide can be quantified using the empirical formula used in the study by Li et al. [3], as shown in Eqs. 2 and 3.

$$R = \left(\prod_{i=1}^{n_s} \xi_i\right)^{1/n_s} \tag{2}$$

where resistance factors are expressed as ξ_i which is a non-dimensional quantity; ξ_i is the ith of $n_s \geq 1$, contributing to the definition of various parameters used in the study.

Hence, the resistance of buildings (R) based on these four parameters can be quantified as follows:

$$R = (\xi_1 \cdot \xi_2 \cdot \xi_3 \cdot \xi_4)^{1/4} \tag{3}$$

5 Case Study

The application of the modified methodology has been illustrated by analyzing physical vulnerability of six buildings in the Chamoli district (Uttarakhand, India). Figure 1 shows an assumed landslide scenario in the Gopeshwar Township of Chamoli district, which is prone to rockfall type of landslide. At first the buildings exposed to rockfall (highlighted in blue color) in the study area were inspected and then the resistance rating is assigned to each indicator based on their severity. Four structural indicators namely structural typology (R1), state of maintenance (R2), construction quality (R3) and number of floors (R4) having resistance factors ξ_1, ξ_2, ξ_3, and ξ_4 respectively were recorded during survey of each building to assess their total resistance. Each

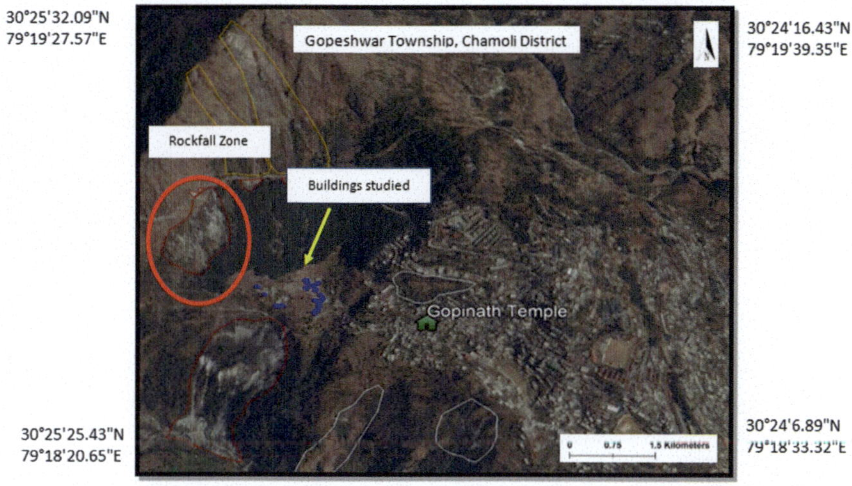

Fig. 1 Study area showing the rockfall zone and the buildings studied (highlighted in blue color) for assessing the physical vulnerability

Table 3 Resistance factor of buildings obtained from the study area

Building ID	ξ_1	ξ_2	ξ_3	ξ_4	Resistance, *R*
1	1	0.8	1	0.5	0.80
2	0.3	0.8	0.5	0.5	0.49
3	0.6	0.2	0.2	1	0.39
4	1	0.8	0.5	1	0.80
5	0.6	0.2	0.2	1	0.39
6	0.6	0.2	0.2	1	0.39

Table 4 Physical vulnerability of the buildings for three different rockfall intensities (PV1—at low rockfall intensity; PV2—at medium rockfall intensity and PV3—at high rockfall intensity)

Building ID	Resistance *R*	PV1	PV2	PV3
I	0.80	0.13	0.88	1
2	0.49	0.33	1	1
3	0.39	0.53	1	1
4	0.80	0.13	0.97	1
5	0.39	0.53	1	1
6	0.39	0.53	1	1

building under study was assigned a building ID and resistance factor corresponding to each parameter and the total resistance of the building, thus obtained is shown in Table 3. The total resistance of the building is calculated using Eq. 3.

Further, the results of physical vulnerability (shown in Table 4) of each building at three different intensities is obtained using Eq. 1.

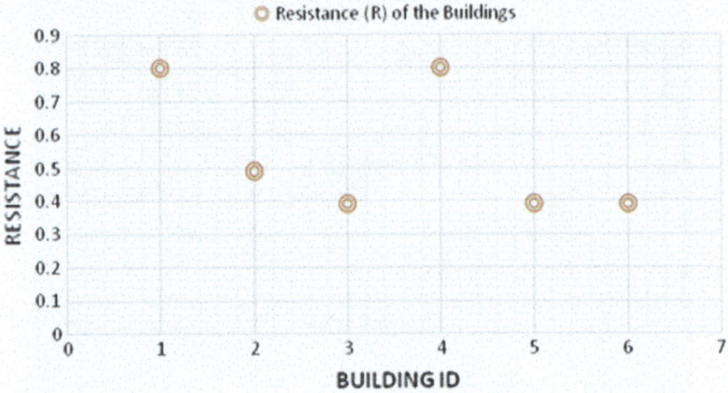

Fig. 2 Resistance of six buildings considered in the study area

6 Conclusion and Discussion

The potential physical loss of buildings when they are affected by large moving masses of debris or rock fragments of some intensity is referred to as physical vulnerability of buildings. Relatively limited methodologies exists based on analytical approach for the quantifying physical vulnerability of buildings (in context to its resistance and intensity) which are subject to actions due to unstable slope and landslides when compared to other components of quantitative landslide vulnerability assessment [19]. In this paper, a framework has been proposed for assessing the physical vulnerability of buildings exposed to three different intensity of rockfall. The resistance of the buildings plays a significant role in reducing the impact of rockfall, since stronger structures (in context to appropriate structural typology, good state of maintenance and construction quality, etc.) will lead to comparatively less damage in case of landslide. Therefore, the proposed framework has been illustrated by studying six buildings in the rockfall zone. The output of the weighted vulnerability ranges from 0 (i.e., no loss) to 1 (i.e., complete loss).

To test the proposed framework six buildings exposed to rockfall zone were analyzed. The resistance of the buildings was determined after field survey and physical vulnerability is quantified at three different intensities. Figure 2 shows the resistance of each building and Fig. 3 shows the physical vulnerability (PV) of buildings exposed to rockfall at three different intensities (PV1: physical vulnerability at low intensity, PV2: physical vulnerability at medium intensity and PV3: physical vulnerability at high intensity). From the results, it is observed that as the rockfall intensity increases, there is an increase in vulnerability of buildings. However, it must be noted that the vulnerability may be reduced if higher resistance is offered by structural indicators to the buildings.

Thus, the vulnerability of buildings in rockfall prone areas provides an objective parameter from the perspective of cost–benefit analysis for the protection interventions. The proposed vulnerability assessment framework is designed to be applied

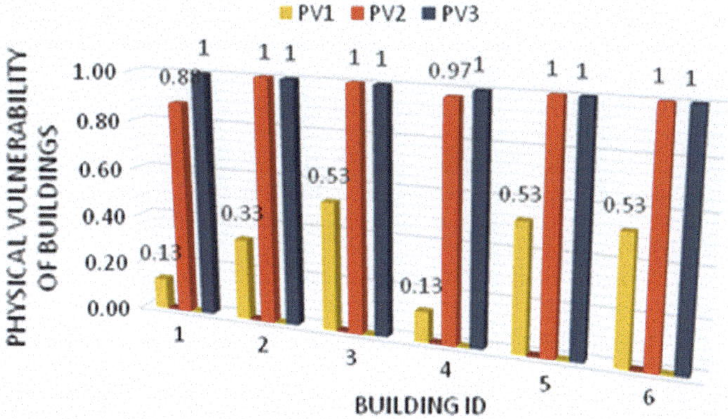

Fig. 3 Plots showing physical vulnerability (PV) of buildings exposed to rockfall at three different intensities (PV1: physical vulnerability at low intensity, PV2: physical vulnerability at medium intensity and PV3: physical vulnerability at high intensity)

on a site-specific scale in areas affected by rockfall activity (from low to moderate) by assessing vulnerability of each and every building (individually), in the case of a single block impact on it. However, in this study only four structural indicators are included, for calculation of resistance. In the future work, more structural parameters need to be included to determine the overall resistance offered by the buildings.

References

1. Glade T, Crozier M (2005) The nature of landslide hazard impact. In: Glade T, Anderson M, Crozier M (eds) Landslide hazard and risk. Wiley, Chichester, pp 43–74
2. Mansour MF, Morgenstern NR, Martin CD (2011) Expected damage from displacement of slow-moving slides. Landslides 7:117–131
3. Li Z, Nadim F, Huang H, Uzielli M, Lacasse S (2010) Quantitative vulnerability estimation for scenario-based landslide hazards. Landslides 7(2):125–134
4. Cruden DM, Varnes DJ (1996) Landslide types and processes. In: Turner AK, Schuster RL (eds) Landslides: investigation and mitigation (Special Report 247). Washington, DC: National Research Council, Transportation and Research Board, pp 36–75
5. Corominas J, Westen C, Frattini P, Cascini L, Malet J-P, Fotopoulou S, Catani F, Eeckhaut O, Mavrouli M, Agliardi F, Pitilakis K, Winter MG, Pastor M, Ferlisi S, Tofani V, Hervás J, Smith T (2014) Recommendations for the quantitative analysis of landslide risk. Bull EngGeol Environ 73(2):209–263
6. Papathoma-Köhle M, Neuhaüser B, Ratzinger K, Wenzel H, Dominey-Howes D (2007) Elements at risk as a framework for assessing the vulnerability of communities to landslides. Nat Hazards Earth SystSci 7:765–779
7. Agliardi F, Crosta GB, Frattini P (2009) Integrating rockfall risk scenario assessment and countermeasure design by 3D modelling techniques. Nat Hazard Earth Sys 9:1059–1073
8. Mavrouli O, Corominas J (2010) Vulnerability of simple reinforced concrete buildings in front of the rockfall impact. Landslides 7(2):169–180
9. Mavrouli O, Corominas J (2010) Rockfall vulnerability assessment for reinforced concrete buildings. Nat Hazard Earth Syst 10:2055–2066

10. Mavrouli O, Fotopoulou S, Pitilakis K, Zuccaro G, Foerster E, Corominas J (2014) Analytical methodologies for the quantification of the vulnerability of buildings to landslides using fragility curves. Bull EngGeol Environ 73:265–289
11. Heinimann HR (1999) Risiko analyse beigravitativen Naturgefahren – Fallbeispiele und Daten. Umwelt-Materialen 107/I, Bern
12. Uzielli M, Nadim F, Lacasse S, Kaynia AM (2008) A conceptual framework for quantitative estimation of physical vulnerability to landslides. EngGeol 102:251–256
13. Dai FC, Lee CF, Nagi YY (2002) Landslide risk assessment and management: an overview. Eng Geol 64:65–87
14. Corominas J, Westen C, Frattini P, Cascini L, Malet J-P, Fotopoulou S, Catani F, Eeckhaut O, Mavrouli M, Agliardi F, Pitilakis K, Winter MG, Pastor M, Ferlisi S, Tofani V, Hervás J, Smith T (2014) Recommendations for the quantitative analysis of landslide risk. Bull Eng Geol Environ 73(2):209–263
15. Corominas J, Copons R, Moya J, Vilaplana JM, Altimir J, Amigó J (2005) Quantitative assessment of the residual risk in a rock fall protected area. Landslides 2:343–357
16. Romana M (2009) Clasificaci´on de losTaludes y LaderasInestablesen Roca enFunci´on de susDimensiones. In: Proceedings of Simposio de taludes, vol 2. Barcelona, Spain, pp 968–979
17. Lateltin O, Hemi C, Raetzo H, Bonnard C (2005) Landslide risk management in Switzerland. Landslides 2:313–320. https://doi.org/10.1007/s10346-005-0018-8
18. EPFL (2002) EcolePolytechnique de Lausanne: relevant criteria to assess vulnerability and risk. Unpublished Deliverable (D16) of project IMIRILAND: impact of large landslides in the mountain environment
19. Castleton J (2009) Rock-fall hazards in Utah, USGS Publications, PI-94
20. Wieczorek GF, Snyder JB (2004) Historical rock falls in Yosemite National Park, US Geological Survey Open-File Report 03-491, http://pubs.usgs.gov/of/2003/of03-491/of03-491.pdf (last access: September 2010)
21. UNDP/UNESCO Vulnerability and Seismic Hazard (1982) Vulnerability analysis in the Balkan region. Project RER/79/014 WG.B Final report
22. Amatruda G, Bonnard C, Castelli M, Forlati F, Giacomelli M, Morelli M, Paro L, Piana F, Pirulli M, Polino R, Prat P, Ramasco M, Scavia C, Bellardone G, Campus S, Durville JL, Poisel R, Preh A, Roth W, Tentschert EH (2004) A key approach: the IMIRILAND project method. In: Bonnard C, Forlati F, Scavia C (eds) Identification and mitigation of large landslide risks in Europe—advances in risk assessment. European Commission Fifth Framework Program. Balkema, Rotterdam, pp 13–44
23. Spence RJS, Kelman I, Calogero E, Toyos G, Baxter PJ, Komorowski JC (2005) Modelling expected physical impacts and human casualties from explosive volcanic eruptions. Nat Hazards Earth SystSci 5:1003–1015
24. Zêzere JL, Garcia RAC, Oliveira SC, Reis E (2008) Probabilistic landslide risk analysis considering direct costs in the area north of Lisbon (Portugal). Geomorphology 94:467–495
25. Silva M, Pereira S (2014) Assessment of physical vulnerability and potential losses of buildings due to shallow slides. Nat Hazards 72:1029–1050

Aditi Singh is a PhD research scholar in Civil Engineering Department at Gautam Buddha University, Greater Noida. She holds a 5 year dual degree B.Tech (Civil Engineering) with M.Tech (Structural Engineering) from the same University. Her research interests include landslide hazard, vulnerability assessment and risk assessment, remote sensing and GIS applications in landslide. She has research publication in reputed international/national journal and conference.

Dr. Shilpa Pal is Assistant professor at Gautam Buddha University, Greater Noida. She holds a B.E. (Civil) from Thapar Institute of Engineering and Technology, M.E. (Structures) from Punjab Engineering College, Chandigarh and PhD from IIT Roorkee. Her research interests include 2-D

& 3-D Seismic Analysis of Dams and Seismic Slope Stability Analysis, Application of remote sensing and GIS in landslides. She has published several publications in international/national journals and conferences.

Dr. D. P. Kanungo is a Senior Principal Scientist in Geotechnical Engineering Group at CSIR-Central Building Research Institute, India. He is also a Professor of Academy of Scientific and Innovative Research (AcSIR), India, in the Faculty of Physical Sciences. He did his Masters of Technology in Applied Geology in 1990 from the University of Roorkee, India (now IIT Roorkee), and PhD in Earth Sciences in 2006 from IIT Roorkee, India. He has professional experience of more than 23 years in the field of engineering geology and landslide disaster mitigation. He has contributed more than 90 research papers in books, journals and conferences. His research interests are engineering geological investigations, landslide hazard and risk assessment, landslide dynamics through instrumentation and monitoring, landslide modelling and remote sensing and GIS applications. He has carried out a large number of infrastructure and landslide disaster mitigation-related projects in India. He is the recipient of Raman Research Fellowship Award in the year 2010 from Council of Scientific and Industrial Research (CSIR), India, and carried out his postdoctoral research at Research Centre on Landslides, Disaster Prevention Research Institute, Kyoto University, Japan.

To Study the Influence of Insurance Policy on the Agriculture Field and Indian Economy: Concept Paper

Yashwant Kumar and Pravin Kumar Singh

Abstract India is an agriculture-dependent country. Agriculture has gone through two major revolutions, the Mechanical revolution, and the Science revolution (often known as Green Revolution). Even then, the system of our agriculture is improper and dotted by corruption. These problems directly affect on our farmers—they lose potential profit and are burdened with debts, leading to suicide, which is counterproductive to our economy, and wastage to our energy resources. This paper introduces a new concept on the central system of agriculture by providing insurance to farmer. In this methodology better utilizes the existing benefits and Indian government's Yojanas will implement to the farmer. The best use of the benefits will depend upon the condition of the particular farmer. Each farmer will facilitate with an insurance policy. Besides of this policy, the various information of the farmer regarding their farming conditions will be stored by the government and further the essential requirement will provide to the greedy farmer. By help of this methodology following benefits will create:

- The farmer's lands and crops would be insured, hence reducing their troubles, and lowering farmer suicide rate.
- An advisory board would help educate the farmers and help deal with problems.
- It would make the government aware of the condition of the nation in terms of the production, and the fertility of the soil of the country.
- The government would be provided with extensive data on the total production of a year, the total land being used for production, and the status and quality of various regions of the nation, along with other statistics.
- Better management would also help in knowing a more accurate investment structure which would be required in the agricultural sector.

Keywords Agriculture · Energy · Insurance · Governance · Agricultural system

Y. Kumar
Department of Civil Engineering, Amity University Jharkhand, Ranchi, India
e-mail: yashwant.bihar2011@gmail.com

P. K. Singh (✉)
Amity University Jharkhand, Ranchi, India
e-mail: pravinsingh.phd@gmail.com

J. Chattopadhyay et al. (eds.), *Renewable Energy and its Innovative Technologies*,
https://doi.org/10.1007/978-981-13-2116-0_2

1 Introduction

Agriculture is the leading sector in terms of employment and livelihood in India, half of Indian employees engaged in farming as the primary occupation. For many industries, agriculture is still a significant contributor of raw materials. The future of Indian agriculture will be the technology-intensive. The current economy shows that, one-fifth of agriculture system in GDP is reduced from one-half of the GDP with respect to the time of Independence of India (1947). Development of modern and advance technologies will be very vital to achieve a higher rate of agricultural growth to avail and fulfill the needs of increasing population. Agriculture is the pillar of India's economy. Rural households of more than 58% depend on the agriculture [1] because it is the main source of income and living. But India is facing various problems in agriculture sector, this happens possibly because of a lack of governance, improper infrastructure, irrigation problems, corrupted middlemen, etc [2].

Indian Government is providing various beneficial Yojana's to the farmers but still the proper contribution of various schemes are not provided to farmers even a proper data on the agricultural sector is not available to the Indian Government, for example the data regarding the area of land being utilized for farming, or how much the final production of crops can be expected is not available and due to this a proper implementation of various schemes is not much success full [3].

1.1 Research in Indian Agriculture

Though research and survey wise in post-independent India, the agriculture confronted many challenges and simultaneously bagged many achievements, even though the daunting challenges are still persisting like low productivity before the huge population (consumer) [4, 5]. Another mounting problem is institutional mismanagement, i.e., even after surplus agro yield, the chase between balanced economy and people's poverty is continuously widening. This shows that the policy-related decisions and its implementation, both are handicapped and thus the farmers are not benefited. In fact the policy reforms should be prone to efficient investment for uplifting rural infrastructure, irrigation, research and development in agriculture, education, and health of rural people.

Right policy and right public investment for enhancing agriculture should be prone to research and development of technological arena for the agro growth in India. Truly saying as the primary sector (agriculture) will develop automatically, the chances of growing tertiary sector will be certain. Under such thoughtful umbrella evidence, say that (TFP) Total Factor Productivity which was the main engine of agricultural growth during 1980s has been showing down today. There are, in few cases, negative growth also like after super investment and research, India moved successfully in Green Revolution (abundant food) but it still has been housing one-fourth of world's hungry and poor mass and 40% of world's malnourished children and women [6]. The latest report of NSSO-2005 reveals that 40% Indian farmers are so acutely threatened in the Agro-Arena, therefore they are deciding to relinquish their job, occupation, or profession of farming [7–9]. Really, its alarming situation for India (Fig. 1).

Undoubtedly agriculture is the largest sector for the livelihood in India, especially in the rural areas. It also contributes a significant figure to the Gross Domestic Product (GDP). Even then, the system of our agriculture is improper. For holistic rural development, India needs to focus on sustainable agriculture which is based on rural employment, food security, environmentally sustainable technologies etc.

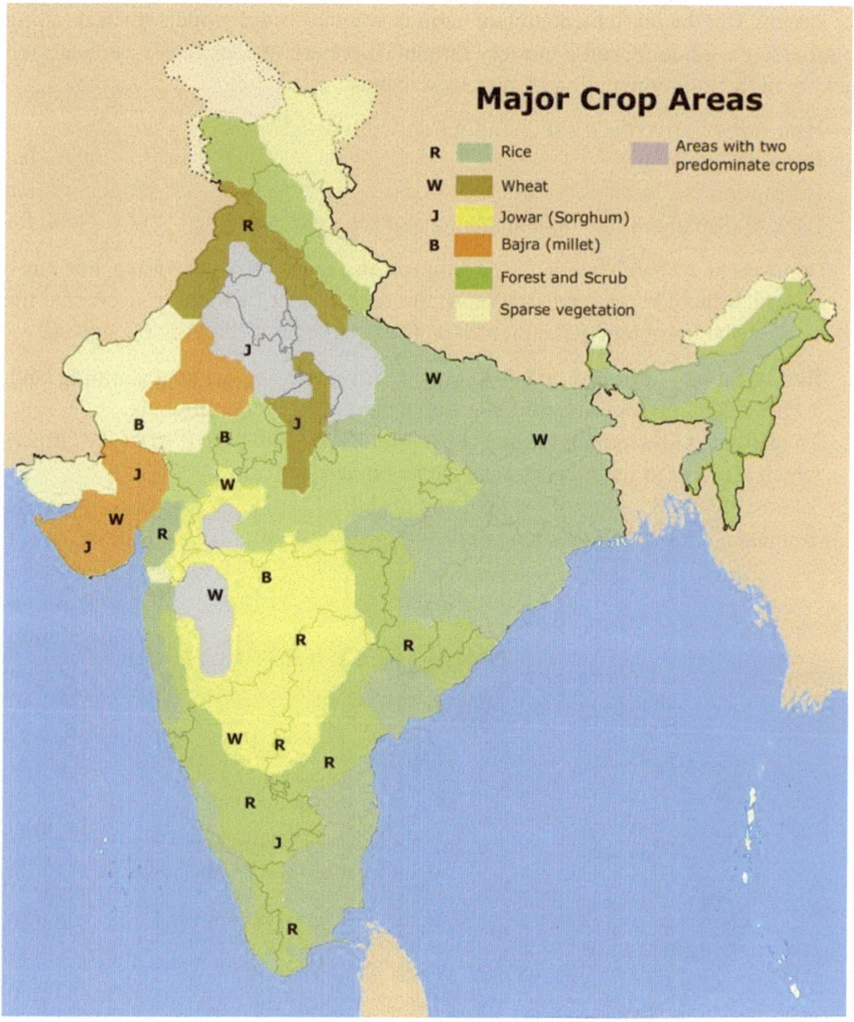

Fig. 1 Agricultural map of India. *Source* Google India [10]

1.2 Agriculture Financing

According to NSSO 70th round report, non-institutional channels are playing a key role in source of credit as nearly 40 and 26% of loans came from informal sources and by moneylenders respectively. Marginal landholding households undergo with only 15% of their loan from institutional sources like government, cooperatives, and banks (Fig. 2).

Day-by-day government is making ease policies to connect with the framers as it is seen by the improvement in terms of institutional coverage. But it need more. For the betterment of agriculture of India government is making lots of Yojna to enhance the system. But the implementation of various schemes is not proper which is failing to provide the actual benefits to every farmer. There are various factors are cause of lacking in the implementation of various schemes:

1. Poor way of advertisement of various agriculture schemes
2. Improper guideline
3. Corruption
4. Lack of data about the implementations of schemes, etc.

The present research work is focusing on the concept of providing insurance policy which includes the registration to the each farmer of India to overcome the above-described problems. The present concept paper has following objectives:

1. Registration of farming Land and farmers card and Insurance to the farming land.
2. To provide Crop security, Soil checkup and making marketing hub.
3. To facilitate Government Kendra at every block level in each district for farmers which are govern by the central and state bodies.
4. To provide employment for skilled and unskilled labors.
5. To make a proper Information sector for 24 h help care.

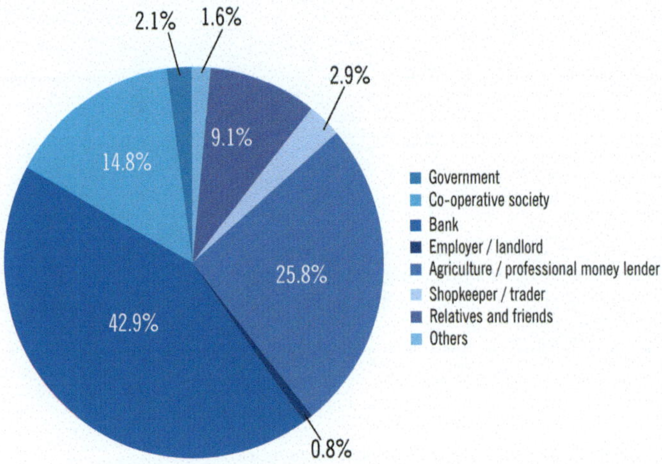

Fig. 2 Loans for agriculture [11]

2 Methodology

Registration of the land provides a Farmer's Card to individual farmer, based on the ownership of the land on which farming has to be done and registered in name of the owner of the land. The registration process will proceed with following steps:

- The Farmer's Card will be linked with the Aadhar Card (of the owner of the land), the mobile number of the owner of the land, an email of the owner of the land (if present), and the owner's respective bank account. So that government can reach to the farmer directly. The registration process will free for the farmers.
- Farmers have to submit each documents related to his/her land, along with few quality of pictures of the empty farmland should be provided. The process of registration should be made such that it can be done offline as well as online.
- After registration process a Farmer Card will be provided to each farmer, these cards contain all the information regarding the farmer's personal detail, farmer's land and other conditions.
- The embedded chip would hold information about the area of the land being used for cultivation, source of irrigation, along with other details mentioned later.
- Insurance policy covers various beneficial aspects for the farmers, it covers all natural disasters—all natural actions which can harm the crops it covers all the natural calamities, fire, accidents, diseases, and such.

  Every season the farmers have to pay the Insurance Charge, these charges will depend on type of crop, geographical condition of location, previous yield data of that land or locality, disaster years in the specified area and the indemnity level of yield crop, and would have to be paid every time when farmers plants a new set of crops.

  The insurance claim would, in the least, cover all the expenses (seeds, fertilizers, labor, etc) provided by the farmer. If the farmer is being funded from a loan then the insurance policy will help to the farmer, at least, to cover the loan redeemed.

As concluded, the purpose of the Farmer's Card is to serve as a verification of identity for the insurance, to help and simplify the process of insuring the lands of the farmers, and to help the process of claiming insurance money.

Figure 3 shows the flow chart of the system which have to be considered for the concept of insurance policy.

In this system, the highest authority is occupied by the central government. Under them falls the state government, under which will be the district. Each district has a block, whose jurisdiction will be managed by the block level Krishi Kendra.

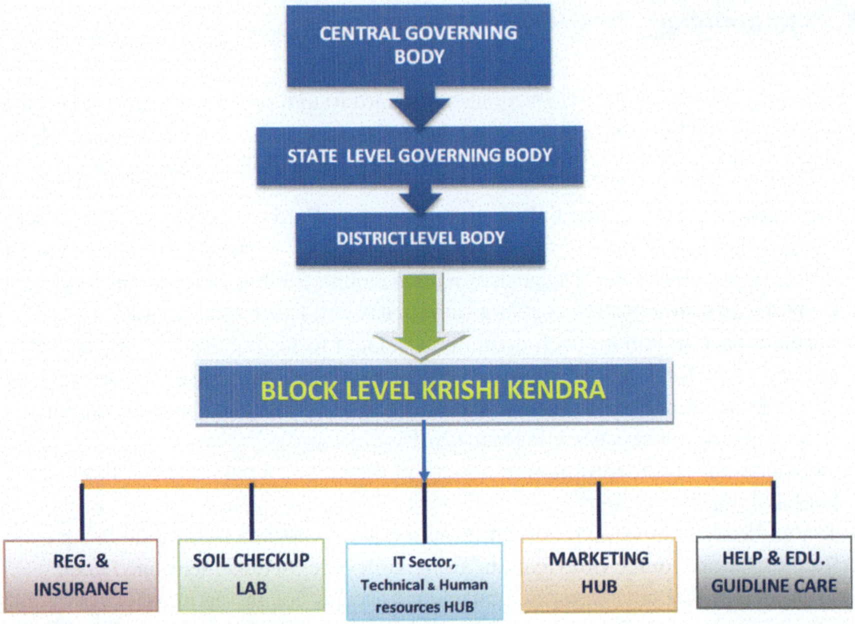

Fig. 3 Flow chart of the system

The Krishi Kendra contains five distinct departments. The first, Registration and Insurance Department, which will be responsible for registration, insurance, and everything related to it. The second is the Soil Checkup Lab Department, which will be responsible for checking the condition and quality of the soil and reporting it to the farmers.

The third is the IT, Technical, and Human Resource department. It will be responsible for educating the farmers about how modern technology can aid them and provide the aid and will also serve as a hub for hiring skilled and unskilled laborers, drivers, machineries and such.

The forth is the Market Hub, which will be responsible for everything related to trading and will allow the farmers to purchase fertilizers and other products.

The fifth is the Help and Educational Guidance Department. It will be responsible for answering any queries the farmers might have and would act as a help desk. It would also help in educating farmers about how to intelligently farm and about modern techniques and technology related to farming.

2.1 Procedural Steps for Insurance Policy

Figure 4 shows the procedural steps to be followed, following are the detail of the steps:

(I) First of all the farmer will have to register with their local Kendrya. During registration they will provide all data about their land. Also, they have to inform about the frequent problems related to agriculture that they used to face during the particular session, also they have to give the common requirements for their agricultural needs, so that the frequent problems will be easily identified by the government. The farmers should also inform the Kendrya about their average household consumption of common crops.
Benefits include

1. Government will have data about amount of land being used for cultivation and an approximate quantity of yield that will be produced.
2. Government will have an idea about the problems being faced on a local level and hence those problems can be dealt with quickly and efficiently.

(II) Before every new plantation, the farmers will have to declare the type/s of crops which will be planted.
Benefits include

1. The government can expect the produce that will be harvested.
2. The government can also prepare for the required fertilizers beforehand.

(III) The farmers would have to provide a sample of their soil to the Kendrya. The Kendrya will test the soil for its nutrient levels and for pH, along with other common tests. This has to be done before every new batch of crops is planted. Also, all the information regarding the soil and land will be saved on the Farmer's Card.
Benefits include

1. It will tell about the requirements, if any, of the land, hence giving a better yield and quality products.
2. It will tell about which crops are more suitable for that specific plot, since most farmers do not have that technical knowledge.

(IV) To help the Kendryas; universities, agricultural colleges, agricultural departments and other such institutions, if available nearby, should help in analyzing the soil samples if the workload is excessive.
Benefits include

1. The students, or the participants, will get research opportunities and extracurricular tasks.
2. The students will get an exposure to the workings of the outside world, will be introduced to new concepts, and will also have the opportunity to learn about and improve the agricultural sector.

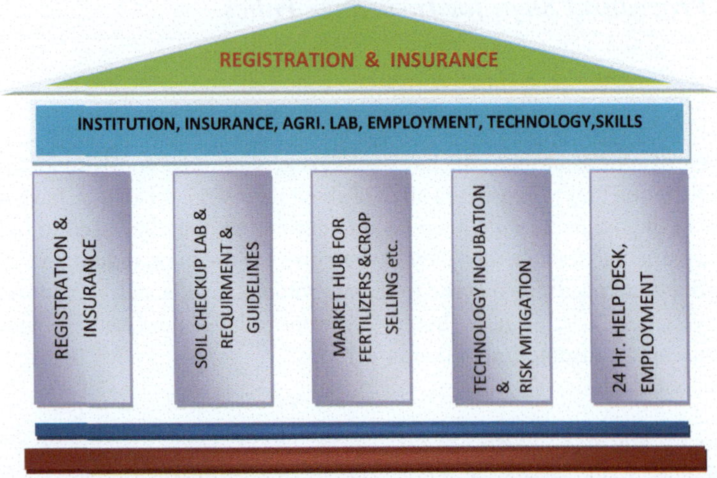

REGISTRATION & INSURANCE	SOIL CHECKUP LAB & REQUIREMENT	MARKET HUB FOR FERTILIZERS & CROP SELLING etc.	TECHNOLOGY INCUBATION & RISK MITIGATION	24 Hr. HELP DESK, EMPLOYMENT
1. To get the actual data of owner & farming land, problems etc. 2. Insurance for crop security.	1. To check the condition of soil. 2. To check which crop will be suitable. 3. To get the approx requirement for fertilizers etc.	1. Registered shop for providing fertilizers etc to farmers. 2. Market hub for selling final production by farmers.	1. Execution of policies related to modern & new technologies in seeds, crop nutrition and protection 2.Promoting new research 3. Development and utilization of IT in field of agriculture 4.More rapid financial inclusion	1. 24hours help facility should be provided to farmers. 2. To provide Skilled & unskilled labour.

Fig. 4 Procedural steps and concepts for insurance policy

(V) The farmer, having obtained all data in a simplified manner, will be suggested fertilizers and other products that should be used to improve the quality of land and soil.
Benefits include

1. It would increase overall produce of the land, and in turn of the nation.
2. It would increase the efficiency of our lands and farmers.

(VI) All the data regarding the suggestions will be stored on the Farmer's Card, and hence the farmer can take the card to a government authorized outlet to purchase products like fertilizers and such.
Benefits include

1. This would reduce corruption in the market, since most fertilizers are sold by corrupt middlemen.
2. Efficiency would be increased since farmers would have data about exact amount of fertilizers required.
3. Since the products will be provided by the government, the quality of the products will have a standard, hence increasing the produce.

(VII) Since all data will be stored on the Farmer's card, it'll help reduce corruption. That is explained in the following points:

1. First, the data submitted by the Kendrya regarding the requirements of the soil and land will be stored online on a database.
2. The outlet providing the products will also be supplied the products in accordance to the demand, which can be coordinated with the help of the database.
3. Since both data is to be equal, the outlet is disabled from selling the products in black; since doing so would show a difference in the database. Hence, corruption would be eliminated.

(VIII) The farmer will have to take a sample of the soil and a sample of the growing crop for verification and for a regular checkup of the quality of soil, hence maintaining the quality of the harvest.
Benefits include

1. The condition of the soil will be known to the Kendrya, hence any discrepancies in the quality of the soil can be dealt with in a timely manner.
2. If the farmer has not mixed the adequate amount of fertilizers, either due to human error or to gain a monetary advantage, the Kendrya will be informed and can take appropriate actions.

(IX) The farmers should inform the Kendrya about the approximate produce that will be harvested around one month before the harvest. This will enable the government to know the total produce of a certain product. Hence having this data would allow the government to better do the following:

1. Have a better data about the quantity of crops that need to be stored in government warehouses, and how much has to be in the market, and how much can be exported for economical benefits.
2. The government can set up a standard rate for different varieties of crops.
3. The government can be pre-informed about which sections will have extra produce and which can face a shortage of crops, hence can better manage distribution of crops on a national scale, which leads to a stable market.

(X) The trading panel, set up in the Kendrya, will be a great asset to the farmers. The farmers will have a central hub where they can trade their crops with the governments. Also, in accord with the aforementioned points, the farmers will have a standard rate at which they will be able to sell their produce. Also, they will be better off with selling at a standard rate. Along with that, the farmers will have an assurance that their produce will be sold to the government, hence their products will not be wasted.

(XI) Skilled and unskilled laborers, and drivers, can register with the Kendrya to offer paid services. Also, people or corporations can register with the Kendrya to offer machineries at a rental cost.
Benefits include

1. This would help drive employment rate up and also prove better for the economy.
2. The farmers would have a central location to hunt for laborers and machineries.
3. The farmers will not have to face unfair prices.
4. Laborers will get a fair price for their work and will have a steady seasonal employment.

3 Conclusion

An introduction of this new system would immeasurably helpful for the farmers. Current problems faced by farmers, like debt, depression, and suicide, could be vastly reduced due to the implementation of this concept. By implementing the central system including quality products and tools, a strong work force will help to drive up the production of crops while also benefiting the economy (Table 1).

Table 1 Concluded the benefits to the farmers by introducing the insurance policy

System	Benefits
1. Registration and insurance	• The farmer's lands and crops would be insured, hence reducing their troubles, and lowering farmer suicide rate • An advisory board would help educate the farmers and help deal with problems • It would make the government aware of the condition of the nation in terms of the production, and the fertility of the soil of the country • The government would be provided with extensive data on the total production of an year, the total land being used for production, and the status and quality of various regions of the nation, along with other statistics • Better management of the agricultural sector would inevitably result in an increase in the GDP of the country • Better management would also help in knowing a more accurate investment structure which would be required in the agricultural sector • Better management would also help lessen wastage of food and crops • It would help the government in better management of import and export of crops • The new system would help boost the economy by providing various jobs in various sectors • It would also provide opportunity for learning and applying skills by students • Nation-wide implementation of this system would require the government to build a better infrastructure for the agricultural sector • A better infrastructure would mean that people would be more inclined towards joining the agricultural sector • It would help build a more direct channel between the farmers and the government, making it easier for the farmers to receive subsidy, and various other government services • It would also help remove the middleman, lessen corruption in the agricultural sector, and help safeguard the farmers from corruption

References

1. Planning Commission. Document of Eleventh Five Year Plan, 2007-12: 1. Inclusive Growth, Vol.1, (Chap. 1, p 1), and 2. Agriculture, Rural Development, Industries, Services and Physical Infrastructure, vol 4 (Chap. 1, p 1 and Chap. 4, p 79). Oxford University Press, New Delhi
2. Hazell P, Ramasamy C (2009) The green revolution: reconsidered. The Johns Hopkins University Press, Baltimore, p 242
3. Cororaton CB (1998) Rates of return to R&D investment in the Philippines. Philippine Institute for Development Studies, Makati City
4. Evenson RE, Pray C, Rosegrant MW (1999) Agricultural research and productivity growth in India, Research Report No. 109, International Food Policy Research Institute, Washington, D.C
5. Kumar P, Anjani K, Mittal S (2004) Total factor productivity of crop sector in the Indo-Gangetic plain of India: sustainability issues revisited. Indian Econ Rev 39(1):169–201
6. Bhushan S (2005) Total factor productivity growth of wheat in India: a malmquist approach. Indian J Agric Econ 60(2):198–210
7. NAAS, Agriculture Sector: Status and Performance, State of Indian Agriculture, National Academy of Agricultural Sciences, New Delhi (2009)
8. Jha D, Kumar S (2006) Research resource allocation in indian agriculture, policy paper 23. National Centre for Agricultural Economics and Policy Research, New Delhi
9. Beintema Nienke, Adhiguru P, Birthal Pratap S, Bawa AK (2008) Agricultural research investment in India, policy brief 27. NCAP, ICAR, New Delhi, p 2008
10. https://en.wikipedia.org/wiki/Agriculture_in_India
11. http://www.tata.com/pdf/Sarthak-Krishi-Yojana.pdf\

Yashwant Kumar is a B.Tech student in Civil Engineering Department from Amity University Jharkhand, India.

Pravin Kumar Singh is Assistant Professor in Mechanical Engineering Department, Amity University Jharkhand. He is pursuing his Ph.D. from National Institute of Technology, Jamshedpur. He received his master's degree in 2011 from Sant Longowal Institute of Engineering and Technology, Punjab. He has good numbers of SCI and SCI Scopus papers in various reputed journals. Recently, one of his papers was published in Springer Journal (JMST-SCI).

Physical Properties of Dewatered Waste Sludge from Textile Processing Wastewater Facility for Construction Material Application

Md. Tousif Rahman and Nadim Khandaker

Abstract Bangladesh is the leading manufacturer of readymade garments in the world. To feed this industry there are about three thousand textile fabrics processing facilities. These facilities treat their generated wastewater by biological treatment. Biological waste stabilization generates substantial bio-solids that are separated out and dewatered before final disposal. One of the dilemmas this vital industry is facing is how to safely dispose the generated dewatered solids. Towards finding a sustainable solution we have systemically characterized the dewatered solid waste from one of the world's largest towel processing facility to see its possible application in the construction sector for safe sequestration. We have looked at properties such as water/moisture content, ash content, specific gravity, particle size analysis, liquid limit, liquidity index, and presence of chromium, cadmium and sodium chloride. The result of analysis showed that heavy metal is not of concern although the sludge has high sodium salt content. The waste sludge has high water content, high volatile fraction, low specific gravity, and low dry density. As a result, it will reduce the unit weight and compressive strength of the clay bricks if added directly as an additive but adding this waste sludge in mix proportion can be a good solution to produce clay bricks and thus finding a sustainable waste sludge disposal method for the textile industry of Bangladesh.

Keywords Textile · Waste activated sludge · Construction material

1 Introduction

In the last decade Bangladesh economy has grown steadily with an average gross product GDP growth of 6% [1]. The economic development is mostly lead by the textile and garments industries. Bangladesh has become the third largest garment

Md. T. Rahman · N. Khandaker (✉)
Department of Civil and Environmental Engineering, North South University, Dhaka 1229, Bangladesh
e-mail: nadim.khandaker@northsouth.edu

J. Chattopadhyay et al. (eds.), *Renewable Energy and its Innovative Technologies*,
https://doi.org/10.1007/978-981-13-2116-0_3

and textile producer and exporter in the world. The garments sector now accounts for about 77% of the country's foreign exchange earnings [2]. But this rapid industrialization is causing some serious environmental problems. One of the major concerns is the disposal of solid waste. Textile mill uses enormous quantity of water and chemical for day to day operations such as washing, bleaching, and dyeing. The wastewater generated from this process cause a significant pollution hazard and makes the treatment of discharged wastewater a priority. The wastewater is treated using both physicochemical and biological process which generates waste sludge. The generated waste sludge has to be disposed of in a benign way that is not harmful to the environment. Landfill is not an option in Bangladesh where there are no classified industrial waste landfills. One disposal idea is to apply the textile generated sludge as construction material additive. Construction industries consume materials that are naturally occurring such as clay, sand, gravel, cement have to be obtained by mining of natural resources with a cost to the environment.

The use of sludge as a construction and building material not only converts waste into useful products but also helps reduce the consumption of natural resources and thus reduce the environmental impact associated with its mining. Many researchers have established that, oven-dried sludge from domestic wastewater treatment plant and sewage ash can be used as an additive in building materials [3–6]. In this paper, we are reporting on an experimental program in characterizing the waste sludge from a wastewater treatment plant that receives wastewater generated from the largest terry towel manufacturing facility in Bangladesh. The treatment plant has a design capacity to treat 6000 m^3 of generated wastewater per day. The plant produces roughly 100 kg/day of waste sludge from its extended aeration activated sludge treatment process. The sludge is thickened, polymer added, and dewatered in a press.

The sludge cake is for the time being stored in house in a walled room. The primary objective is to evaluate physical and chemical parameters of the sludge and compare it to other wastewater treatment sludge published in the literature looking at its application as additive in clay brick production which are then used for construction. The end result is that we can suggest whether this waste sludge from the textile processing wastewater treatment plant can be used as a safe additive in brick making thus finding a benign way to dispose the generated sludge.

2 Materials and Methods

The textile factory produced three tons of waste sludge per month. This sludge usually packed in a 40 kg sack and stored in the waste storage facility. For this research, 300 g of waste sludge was collected from 30 different sacks. From each of these 30 sacks, around 10 g of sample was collected and then mix together to make a representative sample.

The representative sample was analyzed to determine the suitability for application as construction material. A number of tests have been performed: water content, ash content, particle size analysis, specific gravity, density and presence of chromium, cadmium and sodium chloride.

Water content was determined for the waste sludge by at 105 °C in an oven for 24 h. The moisture content test procedure was followed by ASTM D2216 [8]. Hydrometer test for particle size analysis was conducted following ASTM D 422 [9]. Specific Gravity was determined by conducting test following ASTM D 854 [10]. Ash and volatile content was determined by firing the sample in a muffle furnace at 550 °C for 30 min as per Standard Methods, American Public Health Association [11]. Chromium and Cadmium was measured using atomic adsorption spectrophotometer after digestion of the ash fraction of the sludge using 10% nitric acid for 48 h. pH and salinity was measured using HNNA probe and meter.

3 Results and Discussion

Bangladesh a leading manufacturer of readymade garments relies heavily on its textile processing industries to provide their fabrics and yarn. The textile processing is a water-intensive process that generates wastewater which has to be treated before discharge. The wastewater treatment process generates both primary clarified sludge and bio-solids generated from the activated sludge treatment process. The generated sludge has to be disposed of in a benign way. In this experimental program, we defined the characteristics of waste sludge generated from the largest terry towel industry in Bangladesh. The sludge of concern is from there extended aeration system which has been dewatered and stored. Figure 1a shows the representative waste sludge collected from the facility.

We conducted test to determine the water content, specific gravity, particle size and ash content, to find out the properties of the waste sludge. To identify heavy metal and salinity, test was conducted to determine presence of cadmium, chromium, and sodium chloride. Acidity or alkalinity was determined by doing pH test. The

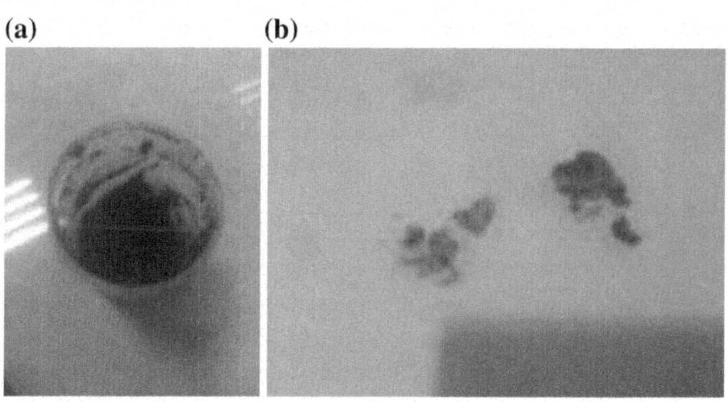

Fig. 1 **a** Picture of representative waste sludge. **b** Picture of waste sludge ash

Table 1 Physical properties of the textile factory generated waste sludge

Property	Value	
	Samples from Terry towel textile factory, Bangladesh	Samples from textile industry CETP from Tamilnadu, India [6]
Color	Dark Greenish	Brown
Appearance	Watered fine solids	Agglomerated fine solids
Specific gravity	1.04–1.06	2.32
Dry density	1040–1060 kg/m^3	2320 kg/m^3
Average particle size	0.018 ± 0.02 mm	0.295 mm
Average water content	88 ± 3.98%	
Volatile fraction	94 ± 4.02%	
Ash	6%	
pH	5.68 ± 0.05	10.5
Cadmium	Not detected	5.6 mg/kg
Chromium	Not detected	358 mg/kg
Sodium chloride	1.9%	

objective was to identify the properties of the waste sludge and to compare the results with similar published research in order to evaluate the possibility of using this waste sludge as an additive in clay brick production. The comparative evaluation is summarized in Table 1.

The representative sample which was collected from the Terry towel factory, Bangladesh was dark greenish in color compare to the brown colored sludge collected from Textile Industry CETP in Tamilnadu, India [6]. Specific gravity and dry density for the representative waste sample was found in the range of 1.04–1.06 and 1040–1060 kg/m^3 whereas waste sludge from Tamilnadu textile CETP was found to be 2.32 and 2320 kg/m^3 respectively [6]. The low specific gravity and dry density of the waste sludge observed if added as an additive in the clay brick production will reduce the unit weight of the material and thus reduce the strength of the brick.

It is also observed that, the representative sludge sample had 88% water content, 94% volatile fraction, and 6% ash content. Which implies that, adding waste sludge as an additive to the clay brick may exceeds the shrinkage limits and thus reduces the compressive strength of produced bricks [6]. So to minimize this effect, Baskar et al. [6] suggested that, the waste sludge can be added in mix proportion to produce quality clay bricks.

It is very significant that, no heavy metal such as cadmium and chromium was detected in the representing waste sludge sample. So, using waste sludge from the representative sample as an additive for clay brick will not have any negative impact in the environment. But substantial amount of sodium chloride was detected in the representative sample which indicates that the sodium chloride residues will cause expansion in the produced brick [7]. According to Ibrahim et al. [7], the presence of sodium chloride underneath the fired clay brick and mortar surface did not cause any large effect but the bonded fired clay brick exposed to sodium sulfate was visibly cracked due to the expansion of mortar joints. In this paper, we did not determine the presence of sodium sulfate in the waste sludge but for the future work it is suggested that, the presence of sodium sulfate need to check before using waste sludge as an additive for clay brick.

4 Conclusions

The results showed that, the waste sludge from terry towel textile industry can be used as an additive to the clay brick, which eventually will help to regenerate the three tons of waste sludge produced by the textile factory. For future scope, it is suggested that more tests be preformed to find out the optimum proportion of waste sludge and clay to produce quality bricks that satisfy the industry standards. Although no heavy metals were present in the sludge, high concentration of sodium chloride in the sludge is of concern and calls for judicious application when mixing with clay to produce bricks.

References

1. Masud Alom Md (2016) Effects on environment and health by Garments factory waste in Narayanganj City, Dhaka. Am J Civ Eng 4(3):80–83
2. European Commission "Guide book for European investors in Bangladesh", sector profiles, Asia Investment Facility, The textile sector in Bangladesh, p 6
3. Tay JH, Show KY, Hong SY, Chien CY, Lee DJ (2002) Potential reuse of wastewater sludge for innovative applications in construction aggregates. Water Sci Technol 50(9):189–196
4. Leda C, Figueiredo lopes Lucena L, Thome Juca J, Soares J, Portela M (2013) Potential uses of sewage sludge in highway construction. J Mater Civ Eng 26(9):04014051
5. Luciana CSH, Carla EH, Miria HMR, Nora DM, Celia RGGT, Rosangela B (2011) Characterization of ceramic bricks incorporated with textile laundry sludge. J Ceram Int 28:951–959
6. Baskar R, Meera Sheriffa Begum KM, Sundaram S (2006) Characterization and reuse of textile effluent treatment plant waste sludge in clay bricks. J Univ Chem Technol Metall 41(4):473–478
7. Ibrahim MHW, Abu Bakar BH, Johari MAM, Jaya RP, Arshad MF (2011) The effect of Na_2SO_4 and NaCl solutions on the moisture movement of fired clay masonry wall. Int J Sustain Constr Eng Technol 2(1)
8. ASTM D 2216. Standard test method for laboratory determination of water (moisture) content of soil and rock by mass
9. ASTM D 422. Standard test method for particle-size analysis of soils
10. ASTM D 854. Standard test methods for specific gravity of soil solids by water pycnometer
11. APH-AWWA-WEF (1998) Standard methods for the examination of water and wastewater, 20th edn. American Public Health Association/American Water Works Association/Water Environment Federation, New York

Md. Tousif Rahman is a Lecturer in Department of Civil and Environmental Engineering, North South University, Dhaka, Bangladesh. He has completed his master's degree from University of Newcastle (UK). His research area of interest is geotechnical engineering, construction material, slope stability and underground construction.

Dr. Nadim Khandekar is an Associate Professor in Department of Civil and Environmental Engineering, North South University, Dhaka, Bangladesh. He has completed his Ph. D from Pennsylvania State University, USA. He is an expert in translational research where he takes concepts derived in the bench to fruitful industrial application. Dr. Khandaker is a licensed professional engineer in Ontario Canada.

Multilevel Inverter with Optimal Reduction of Power Semi-conductor Switches

Bidyut Mahato, Sudhanshu Mittal, Saikat Majumdar, Kartick Chandra Jana and Paresh Kumar Nayak

Abstract Multilevel Converters is the most emerging topic of research now-a-days due to its enormous various industrial applications like cement factory, ceramic industry, aerospace applications, marine applications and many more. Besides this multilevel inverter enables the usage of renewable energy sources i.e. wind, photovoltaic, fuel cells that can be easily interlinked to a multilevel converter system for the higher power rating applications. In this paper, a novel 9-level and 17-level single phase multilevel inverter has been designed, analyzed, described, and verified in MATLAB/Simulink environment. The prototype model is developed in laboratory and the simulated results of 9-level inverter is confirmed by the corresponding experimental results. In this paper, the comparison chart between the proposed topology and some recent proposed topologies (both the symmetrical and asymmetrical topologies) has been included as well in relations of the total number of power semi-conductor devices, driver circuits, DC supply sources/capacitors and the total components required.

Keywords Power electronic switches · Power converter · 9-Level inverter
Pulse width modulation

1 Introduction

Due to the energy shortage, the integration of renewable energy sources to the electricity grid becomes an interesting research topic nowadays. The number of renewable energy sources and distributed generators is increasing very fast which also brings some threats to the power grid [1]. In order to maintain or even to improve the power supply reliability and quality of the power system with distributed generation, it is necessary to have some new strategies for the operation and management

B. Mahato (✉) · S. Mittal · S. Majumdar · K. C. Jana · P. K. Nayak
Department of Electrical Engineering, Indian Institute of Technology (Indian School of Mines), Dhanbad, India
e-mail: bidyut1990@gmail.com

J. Chattopadhyay et al. (eds.), *Renewable Energy and its Innovative Technologies*,
https://doi.org/10.1007/978-981-13-2116-0_4

of the electricity grid. The multilevel inverter (MLI) output power quality improves with more voltage level with lesser total harmonic distortion (THD) thus being used in various applications [2]. MLI has massive benefits above a classical two-level converter that uses high switching frequency pulse-width modulation (PWM) such as staircase waveform quality, common-mode voltage (CMV), switching frequency and input current. The current rating or the voltage rating of the multilevel inverter becomes a compound of specific switches and thus the power rating of the inverter can surpass the limit executed by the individual switches [3]. There are three universally employed topologies for the multilevel inverter called diode-clamped inverter or neutral-point clamped, capacitor-clamped or flying capacitor, and cascaded multilevel inverter [4–6]. Various applications for different power ranges has been reported in [7–9] and amongst these topologies, the cascaded H-bridges (CHB) multilevel inverter is appreciated as the supreme topology for the renewable energy as the separate DC sources that it needs can be easily fed by the PV arrays, fuel cells or wind turbines [10, 11]. In addition, in order to compensate the fluctuation of the output power of the PV arrays and wind turbines, several energy storage devices can also be incorporated into the system. Furthermore, since the number of DC sources can be chosen arbitrarily, it is convenient to increase the level of the output voltage and output power. To increase the voltage levels, simply the power switches is increased along with the DC sources/capacitors, gate driver circuits makes the circuit more complex and costlier [12]. So, reducing the power switches becomes a great challenge for the multilevel inverters for the researchers. Various recent reduced switch topologies [13–27] have been compared with the topology proposed in this paper. Different symmetrical topologies [13–25] as well as asymmetrical topologies [13, 19, 20, 26, 27] are discussed in terms of number of employed power switches.

A novel reduced switch multilevel inverter is proposed that can be designed for any desired levels with optimum number of required switches. The proposed topology can be designed for both symmetric and asymmetric values of DC sources/capacitors. A 9-level/17-level symmetric has been designed and tested in MATLAB. A hardware set-up is designed in laboratory and the simulation results are verified by corresponding experimental results. Level shifted pulse width modulation is employed to generate switching pulses and the real-time pulses is fed to power switches using the real time simulator dspace1103. In addition, various recent proposed symmetric/asymmetric voltage topologies are compared correspondingly with proposed symmetric/asymmetric topology.

2 Reduced Switch Proposed Structure

The generalized circuit diagram of the proposed single phase multilevel inverter is depicted in Fig. 1. Integrated gate bi-polar transistor (IGBTs) are the power semiconductor switches that have been employed in the overall circuit topology. There are only two unidirectional switches named S_5 and S_4 whereas all the other remaining switches used in the topology are bi-directional switches. S_2 is the switch being

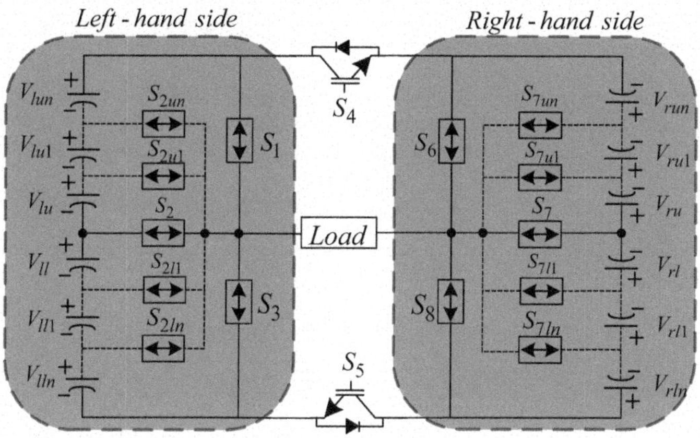

Fig. 1 Generalized structure of proposed topology

connected at left-hand side between the middle point of switches S_1 and S_3 whereas S_7 is the switch being connected at the middle point of right-hand side between the switches S_6 and S_8. S_1, S_3 and S_6, S_8 are the left-hand side and right-hand switches respectively that remains fixed with their positions throughout the generalized topology in spite of increasing voltage levels. S_4 and S_5 are the switches that remains fixed at the top-most point and the bottom-most point for any desirable voltage output levels. Switch S_2 is further bifurcated as upper and lower switches where S_{2un} indicates the *n*th upper-switch and S_{2ln} indicates the *n*th lower-switch as extended on both the upper and lower sides of switch S_2. Switch S_{7un} represents the *n*th upper-switch and switch S_{7ln} represents the *n*th lower-switch as extended on both the upper and lower sides of switch S_7.

One upper switch (S_{2u1}) with one lower switch (S_{2l1}) of left-hand side and one upper switch (S_{7u1}) with one lower switch (S_{7l1}) of right-hand side is required to obtain the next output voltage level. Several capacitors are needed to achieve the stepped output voltage. The voltage across the capacitors are also differentiated as the left-hand side (V_l) and the right-hand side (V_r) of the proposed circuit topology. Capacitor Voltages, V_{lu} (left-hand upper) & V_{ll} (left-hand lower) are the capacitors voltage on the left-hand side whereas V_{ru} (right-hand upper) & V_{rl} (right-hand lower) are the capacitors voltage on the right-hand side are the minimum number of capacitor voltages required to obtain a minimum number of voltage levels in the proposed configuration as depicted in Fig. 1. V_{lun}, V_{lln} and V_{run}, V_{rln} are the *n*th capacitors voltage of left-hand upper, left-hand lower side as well as right-hand upper and right-hand lower side of the proposed circuit topology respectively.

In the proposed topology, both the symmetrical and asymmetrical configurations are feasible depending on the capacitor voltage ratio. On the left-hand side, all the input capacitor voltages (both the upper-side as well as the lower-side) can be fed by a DC source. Similarly, the input capacitor voltages on right-hand side (both upper-side as well as the lower-side) can be fed by a DC source. Therefore, only two

Fig. 2 Single phase proposed symmetric/asymmetric MLI structure

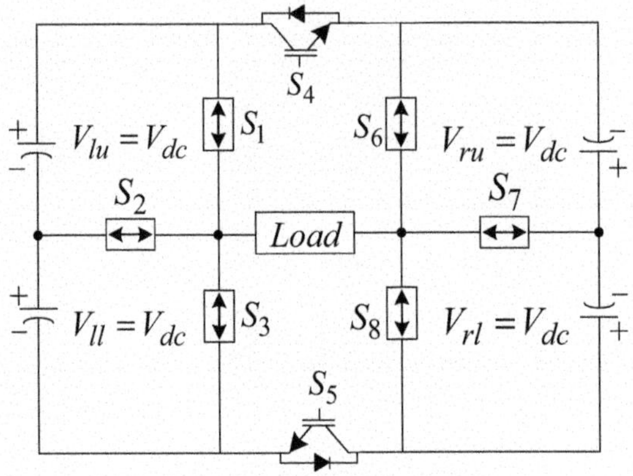

(a) 9-level symmetrical inverter.

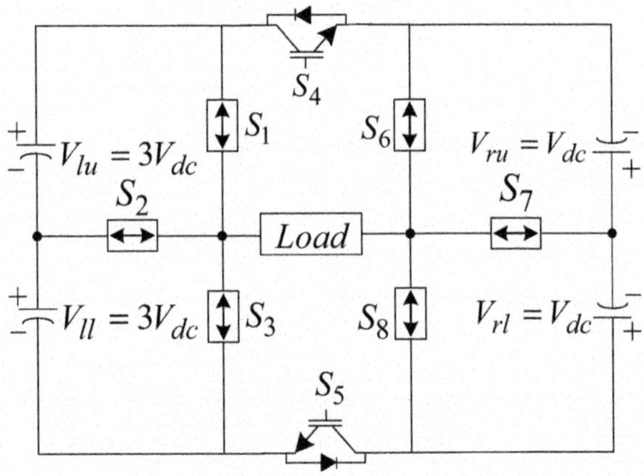

(b) 17-level asymmetrical inverter.

input supply is required (one for left-hand side and other for the right-hand side) to synthesize the stepped staircase voltage waveforms.

Figure 2a, b shows the proposed symmetrical and asymmetrical configuration that has been well explained elaborately. In Fig. 2a, proposed symmetrical configuration having all the capacitor voltages starting from the left-hand upper side (V_{lu}, $V_{lu\,1} \ldots V_{lun}$), left-hand lower-side voltages (V_{ll}, $V_{ll1} \ldots V_{lln}$) and the capacitor voltages starting from the right hand upper-side (V_{ru}, $V_{ru1} \ldots V_{run}$), right hand upper-side (V_{rl}, $V_{rl1} \ldots V_{rln}$) will be of same voltage magnitude as shown below in Eq. (1),

$$\begin{aligned} V_{lu}, V_{lu1} \ldots V_{lun} &= V_{ll}, V_{ll1} \ldots V_{lln} = V_{dc} \\ V_{ru}, V_{ru1} \ldots V_{run} &= V_{rl}, V_{rl1} \ldots V_{rln} = V_{dc} \end{aligned} \tag{1}$$

Fig. 3 **a** Proposed 49-level inverter. **b** Proposed 97-level inverter

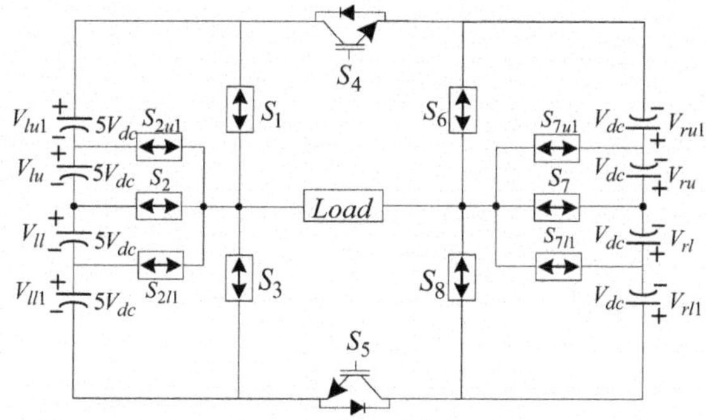

(a) Proposed 49-level Inverter

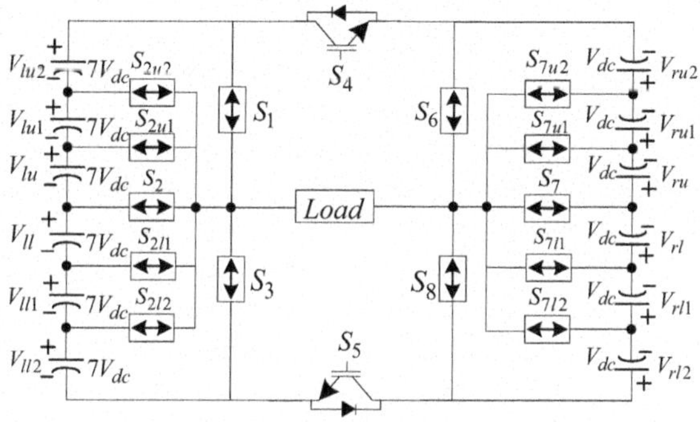

(b) Proposed 97-level Inverter

For the asymmetrical configuration as shown in Fig. 2b, magnitudes of the capacitor voltages will be of different magnitudes. There will be two different voltage magnitudes on either two sides (left-hand side and right-hand side) of described topology. The optimal value of capacitor voltage magnitude for the asymmetrical proposed configuration is shown below as in Eq. (2),

$$
\begin{aligned}
V_{\mathrm{lu}} &= V_{\mathrm{ll}} = 3V_{\mathrm{dc}} \\
V_{\mathrm{ru}} &= V_{\mathrm{rl}} = V_{\mathrm{dc}}
\end{aligned}
\tag{2}
$$

The next achievable voltage levels is 49-levels using 10 bi-directional and 2 uni-directional switches as depicted in Fig. 3a. Similarly, the next achievable voltage levels is 97-levels using 14 bi-directional and 2 uni-directional switches as depicted in Fig. 3b. Therefore, it can be concluded that to obtain the maximum voltage levels in asymmetrical configuration, the value of voltage magnitude on one side (either left-hand or right-hand) of the shown topology will depend on the number of increase of power switches as mentioned above. The magnitude of all the voltage sources on the right-hand side will be V_{dc} whereas the magnitude of all the voltage sources on the left-hand side will be the total sum of the voltage sources on the right-hand side ($5V_{\mathrm{dc}}$ for 49-level and $7V_{\mathrm{dc}}$ for 97-level).

For a symmetrical topology, the minimum number of voltage levels obtained is 9-level. In terms of power semi-conductor switches (unidirectional or bi-directional), there are overall eight switches (S_1, S_2, S_3, S_4, S_5, S_6, S_7, S_8) being involved to generate a 9-level inverter where three of them i.e. S_1, S_2, S_3 are employed on left-hand side and S_6, S_7, S_8 are used on right-hand side of the proposed topology. In terms of employed capacitor voltages, there are overall four capacitors having voltage magnitudes as V_{lu}, V_{ll}, V_{ru}, V_{rl} being involved to generate a 9-level inverter where two of capacitors having voltage magnitudes as V_{lu}, V_{ll} are employed on left-hand side and two of capacitors having voltage magnitudes as V_{ru}, V_{rl} are used on right-hand side respectively. Thereby, an enhancement in required number of switches as well capacitors to achieve higher voltage levels will not affect the connecting terminals (connecting positions) of the overall eight switches (S_1, S_2, S_3, S_4, S_5, S_6, S_7, S_8) and overall four voltage sources (V_{lu}, V_{ll}, V_{ru}, V_{rl}).

To obtain the next output voltage level, both the parameters i.e. switches and the voltage sources required to be increase in some fixed ratio. For example, 17-level is the next output voltage level obtained in asymmetrical configuration that requires the addition of four switches i.e. S_{2u1}, S_{2l1} in left-hand side and S_{7u1}, S_{7l1} in the right-hand side along with the four capacitors having voltage magnitudes as V_{lu1}, V_{ll1} in left-hand side and V_{ru1}, V_{rl1} in the right-hand side. Thus, it is observed that for both the symmetrical and asymmetrical proposed configuration, four switches and four input DC voltage sources is required additionally to the previous proposed topology to achieve next output voltage level. Minimum number of voltage levels obtained in an asymmetrical topology is 17-level and the voltage level obtained next to 17-level is 49-level and so on. The calculation of the voltage levels has been analyzed and shown in Sect. 4.

3 Modes of Operation

The control strategy of the proposed topology is implemented using pulse width modulation technique where carrier signals and a sine signal is taken as reference signals are compared to generate the switching pulses for the power switches employed in the configuration. Among various modulation techniques such as phase shifted PWM [28], nearest level control (NLC) [29], space vector PWM (SVPWM) [30] the level shifted PWM (LS-PWM) is used to develop the pulses in MATLAB and is fed to the real-time simulator dSPACE 1103. Generation of voltage steps at different modes of 9-level inverter has been described below as well. Nevertheless, switching table and switching pulses for 9-level inverter (symmetric voltage ratio) as well as 17-level inverter (Asymmetric voltage ratio) has been incorporated as well. All the modes generating different voltage levels are depicted in Fig. 4 and explained as below.

Mode 1: In this mode, two bi-directional switches S_3, S_8 and one unidirectional switch S_4 are switched on along with all the capacitor voltages of left-hand side (V_{lu}, V_{ll}) and right-hand side (V_{ru}, V_{rl}) to generate $+4V_{\text{dc}}$.

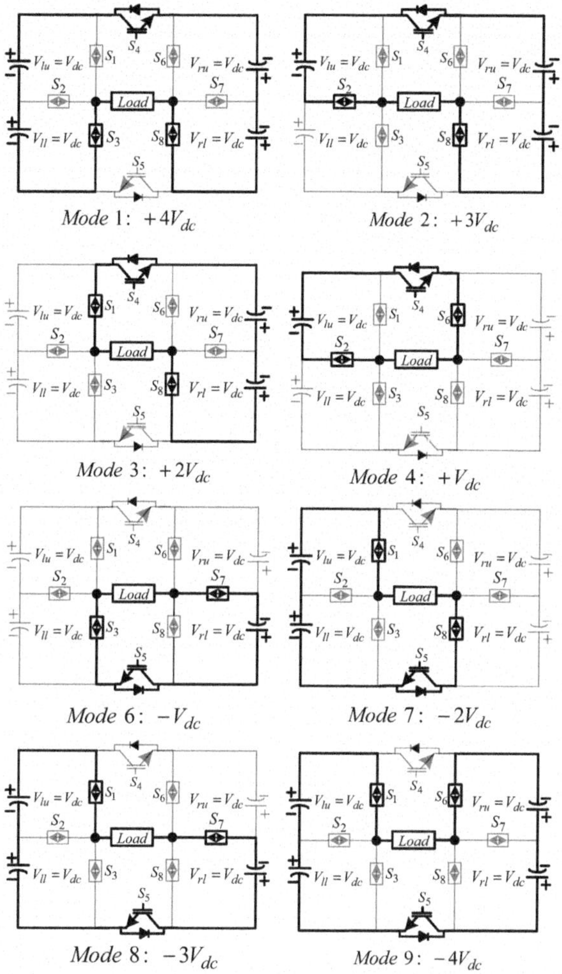

Fig. 4 Various modes with different voltage level generation

Mode 2: In this mode, two bi-directional switches S_2, S_8 and one unidirectional switch S_4 are switched on with the capacitor voltage of left-hand side (V_{lu}) and right-hand side (V_{ru}, V_{rl}) to generate $+3V_{\mathrm{dc}}$.
Mode 3: In this mode, two bi-directional switches S_1, S_8 and one unidirectional switch S_4 are switched on along all the capacitor voltages of right-hand side (V_{ru}, V_{rl}) to generate $+2V_{\mathrm{dc}}$.
Mode 4: In this mode, two bi-directional switches S_2, S_6 and one unidirectional switch S_4 are switched on along with the left-hand side capacitor voltage (V_{lu}) to generate $+V_{\mathrm{dc}}$.
Mode 5: In this mode, two bi-directional switches S_1, S_6 and one unidirectional switch S_4 are switched to generate 'zero' V_{dc}. This mode has not been shown in Fig. 3.

Table 1 Switching table for proposed 9-level inverter

Modes	Conducting switches 1 = ON; 0 = OFF								Output phase voltage $V_{lu}, V_{ll} = V_{dc}$ $V_{ru}, V_{rl} = V_{dc}$
	S_1	S_2	S_3	S_4	S_5	S_6	S_7	S_8	
1	0	0	1	1	0	0	0	1	$+4V_{dc}$
2	0	1	0	1	0	0	0	1	$+3V_{dc}$
3	1	0	0	1	0	0	0	1	$+2V_{dc}$
4	0	1	0	1	0	1	0	0	$+V_{dc}$
5	1	0	0	1	0	1	0	0	0
6	0	0	1	0	1	0	1	0	$-V_{dc}$
7	1	0	0	0	1	0	0	1	$-2V_{dc}$
8	1	0	0	0	1	0	1	0	$-3V_{dc}$
9	1	0	0	0	1	1	0	0	$-4V_{dc}$

Mode 6: In this mode, two bi-directional switches S_3, S_7 and one unidirectional switch S_5 are switched on along with the right-hand side capacitor voltage (V_{rl}) to generate $-V_{dc}$.
Mode 7: In this mode, two bi-directional switches S_1, S_8 and one unidirectional switch S_5 are switched on along with the right-hand side capacitor voltages (V_{lu}, V_{ll}) to generate $-2V_{dc}$.
Mode 8: In this mode, two bi-directional switches S_1, S_7 with one unidirectional switch S_5 are switched on along with the capacitor voltage of left-hand side (V_{lu}, V_{ll}) and right-hand side (V_{rl}) to generate $-3V_{dc}$.
Mode 9: In this mode, two bi-directional switches S_1, S_6 and one unidirectional switch S_5 are switched on along with all the capacitor voltages of left-hand side (V_{lu}, V_{ll}) and right-hand side (V_{ru}, V_{rl}) to generate $-4V_{dc}$.

It can be easily noticed that Tables 1 and 2 shows the conducting switches for the different modes of proposed symmetrical 9-level and asymmetrical 17-level inverter. Though, the number of switches are same for both the cases, number of levels obtained is more in asymmetric magnitude voltage distribution of capacitors rather than the symmetric magnitude voltage distribution of capacitors. Similarly, the pulse pattern or the nature of gate pulses found in Figs. 5 and 6 for the proposed symmetrical 9-level and asymmetrical 17-level inverter are different for the same power semi-conductor switches in either of the cases. All the switching pulse pattern is added from the simulated model tested in MATLAB/Simulink environment.

Table 2 Switching table for proposed 17-level inverter

Modes	Conducting switches 1=ON; 0=OFF								Output phase voltage V_{lu}, $V_{ll}=3V_{dc}$ V_{ru}, $V_{rl}=V_{dc}$
	S_1	S_2	S_3	S_4	S_5	S_6	S_7	S_8	
1	0	0	1	1	0	0	0	1	$+8V_{dc}$
2	0	1	0	1	0	0	0	1	$+7V_{dc}$
3	1	0	0	1	0	0	0	1	$+6V_{dc}$
4	0	0	1	1	0	0	1	0	$+5V_{dc}$
5	0	1	0	1	0	0	1	0	$+4V_{dc}$
6	1	0	0	1	0	0	1	0	$+3V_{dc}$
7	0	0	1	1	0	1	0	0	$+2V_{dc}$
8	0	1	0	1	0	1	0	0	$+V_{dc}$
9	0	0	1	0	1	0	0	1	0
10	0	1	0	1	0	1	0	0	$-V_{dc}$
11	1	0	0	0	1	0	0	1	$-2V_{dc}$
12	0	0	1	0	1	0	1	0	$-3V_{dc}$
13	0	1	0	0	1	0	1	0	$-4V_{dc}$
14	1	0	0	0	1	0	1	0	$-5V_{dc}$
15	0	0	1	0	1	1	0	0	$-6V_{dc}$
16	0	1	0	0	1	1	0	0	$-7V_{dc}$
17	1	0	0	0	1	1	0	0	$-8V_{dc}$

4 Comparison with Other Topologies

Various recent reduced switch topologies proposed in topologies [13–27] have been studied in terms of total required number of switches (N_{sw}),driver circuits, DC supply sources/capacitors, main diodes, clamping diodes, unidirectional switches, bidirectional switches, total components count and output voltage levels. All the aforementioned parameters are presented in Table 3 in terms of total number of switches (N_{sw}) for comparing the effectiveness of the proposed inverter with some existing topologies proposed in recent years by researchers throughout the globe. The generalized formulae are presented to make the calculations of the mentioned parameters simple for any number of required switches. Nevertheless, the proposed symmetrical structure are compared with the existed symmetrical reduced switch topologies and the proposed asymmetrical MLI are compared with the existed asymmetrical reduced switch topologies in Tables 3 and 4 respectively.

The proposed 9-level inverter is further verified by plotting the graph between the required number of power switches and the corresponding voltage levels generated as shown in Fig. 7. Moreover, Fig. 7a indicates the graph between the required number of power switches and the corresponding voltage levels for symmetrical topology whereas Fig. 7b shows the graph between the required number of power switches and the corresponding voltage levels for asymmetrical topology. It is clearly

Table 3 Generalized formulae of proposed symmetrical MLI structure with the other symmetrical topologies

Cited papers	Total number of switches	Driver circuits	Capacitors/isolated DC sources	Main diodes	Clamping diodes	Output voltage levels	Switches		Total components count
							Uni-directional switches	Bi-directional switches	
[13]	N_{sw}	N_{sw}	$\frac{(N_{sw}-4)}{2}$	N_{sw}	–	$(N_{sw}-3)$	N_{sw}	–	$\frac{(7N_{sw}-4)}{2}$
[14]	N_{sw}	N_{sw}	$(N_{sw}/3)$	N_{sw}	–	$2(N_{sw}/3)+1$	N_{sw}	–	$(10N_{sw}/3)$
[15]	N_{sw}	N_{sw}	$(N_{sw}/4)$	N_{sw}	–	$\frac{(N_{sw}+2)}{2}$	N_{sw}	–	$(13N_{sw}/4)$
[16]	N_{sw}	N_{sw}	$\frac{(N_{sw}-2)}{2}$	N_{sw}	–	$(N_{sw}-1)$	N_{sw}	–	$\frac{(7N_{sw}-2)}{2}$
[17]	N_{sw}	N_{sw}	$\frac{(N_{sw}-4)}{2}$	N_{sw}	–	$(N_{sw}-1)$	N_{sw}	–	$\frac{(7N_{sw}-4)}{2}$
[18]	N_{sw}	N_{sw}	$\frac{(N_{sw}-1)}{3}$	N_{sw}	–	$\frac{(2N_{sw}+1)}{3}$	N_{sw}	–	$\frac{(10N_{sw}-1)}{3}$
[19]	N_{sw}	N_{sw}	$\frac{(N_{sw}-4)}{2}$	N_{sw}	–	$(N_{sw}-3)$	N_{sw}	–	$\frac{(7N_{sw}-4)}{2}$
[20]	N_{sw}	N_{sw}	$(N_{sw}-3)$	$(N_{sw}+3)$	–	$(2N_{sw}-7)$	6	$(N_{sw}-6)$	$4N_{sw}$
[21]	N_{sw}	N_{sw}	$\frac{(N_{sw}-4)}{2}$	N_{sw}	–	$(N_{sw}-3)$	N_{sw}	–	$\frac{(7N_{sw}-4)}{2}$
[22]	N_{sw}	N_{sw}	$\frac{(N_{sw}-2)}{2}$	N_{sw}	–	$(N_{sw}-1)$	N_{sw}	–	$\frac{(7N_{sw}-2)}{2}$
[23]	N_{sw}	N_{sw}	$\frac{(N_{sw}-2)}{2}$	N_{sw}	–	$(N_{sw}-1)$	N_{sw}	–	$\frac{(7N_{sw}-2)}{2}$
[24]	N_{sw}	N_{sw}	$\frac{(N_{sw}-2)}{2}$	N_{sw}	–	$(N_{sw}-1)$	N_{sw}	–	$\frac{(7N_{sw}-2)}{2}$
[25]	N_{sw}	N_{sw}	$\frac{3N_{sw}-8}{4}$	N_{sw}	–	$\frac{3(N_{sw}-2)}{2}$	N_{sw}	–	$\frac{15N_{sw}-8}{4}$
Proposed topology	N_{sw}	N_{sw}	$(N_{sw}-4)$	$(4N_{sw}-6)$	–	$2(N_{sw}-4)+1$	2	$(N_{sw}-2)$	$(7N_{sw}-9)$

Table 4 Generalized formulae of proposed asymmetrical MLI structure with the other asymmetrical topologies

Papers	Total switches	Gate driver circuits	Capacitors/isolated DC sources	Main diodes	Output voltage levels	Switches		Total components count
						Uni-directional Switches	Bi-directional switches	
[13]	N_{sw}	N_{sw}	$\frac{(N_{sw}-4)}{2}$	N_{sw}	$(2N_{sw}-9)$	N_{sw}	–	$\frac{(7N_{sw}-4)}{2}$
[26]	N_{sw}	N_{sw}	$\frac{(N_{sw}-2)}{3}$	N_{sw}	$(2N_{sw}-7)$	N_{sw}	–	$\frac{(10N_{sw}-2)}{3}$
[27]	N_{sw}	N_{sw}	$(N_{sw}/2)$	$5N_{sw}/4$	$\frac{(3N_{sw}+2)}{2}$	$3N_{sw}/4$	$N_{sw}/4$	$15N_{sw}/4$
[19]	N_{sw}	N_{sw}	$\frac{(N_{sw}-4)}{2}$	N_{sw}	$\left[\frac{1}{2}(N_{sw}-4)(p+N_{sw})+1\right]$	N_{sw}	–	$\frac{(7N_{sw}-4)}{2}$
[20]	N_{sw}	N_{sw}	$4N_{sw}/7$	$(N_{sw}+3)$	$(8/7)N_{sw}+1$	$6N_{sw}/7$	$N_{sw}/7$	$\frac{(25N_{sw}+21)}{7}$
Proposed topology	N_{sw}	N_{sw}	$(N_{sw}-4)$	$(4N_{sw}-6)$	$(4N_{sw}-15)$	2	$(N_{sw}-2)$	$(7N_{sw}-9)$

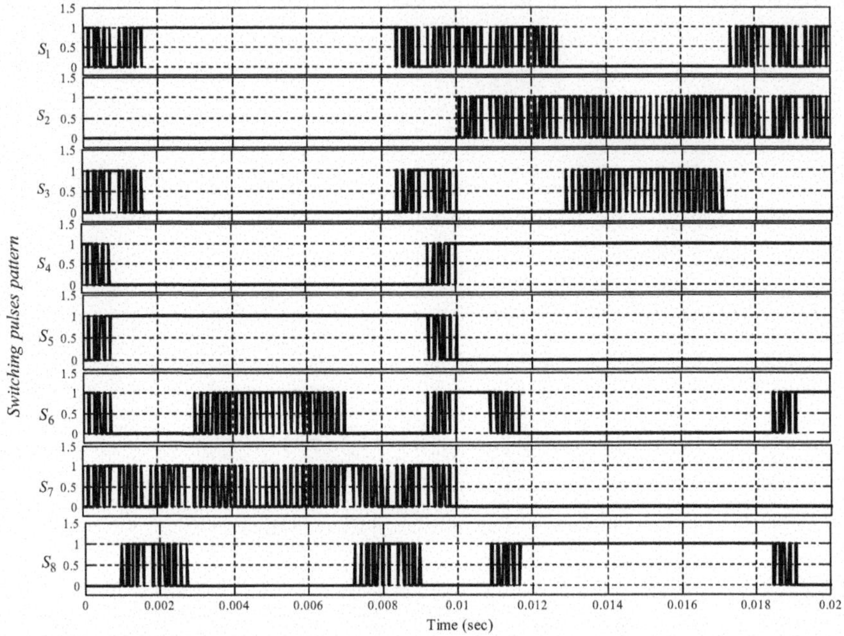

Fig. 5 Pulse pattern of each power switches in proposed 9-level inverter

understood from both the graphs (Fig. 7a, b) that the proposed topology founds better for both the aspects i.e. symmetrical as well asymmetrical topology. Table 5 shows the various stages of the proposed symmetrical as well as asymmetrical topologies. As mentioned in Table 5, stage-I comprises of 8 power switches (2 unidirectional and 6 bi-directional) generating 9-level for symmetrical configuration and the same number of required components generating 17-level for asymmetrical configuration (DC voltage ratio values). Stage–II implies the required number of power switches next to previous required power switches. 12 power switches (2 unidirectional and 10 bi-directional) and 8 DC sources are needed for generating next achievable voltage levels thereby producing 17-level for symmetrical and 49-level for asymmetrical topology and thus all the required number of components and their corresponding voltage levels can be easily calculated from the generalized equation provided in Table 3 (for symmetrical topologies) and Table 4 (for asymmetrical topologies).

Table 5 Various components required for proposed symmetrical and asymmetrical MLI at different stages

	Proposed symmetrical MLI				Proposed asymmetrical MLI			
	Bi-directional switches	Uni-directional switches	Capacitors/isolated DC sources	Output voltage levels	Bi-directional switches	Uni-directional switches	Capacitors/isolated DC sources	Output voltage levels
Stage-I ($N_{sw} = 8$)	6	2	4	9	6	2	4	17
Stage-II ($N_{sw} = 12$)	10	2	8	17	10	2	8	49
Stage-III ($N_{sw} = 16$)	14	2	12	25	14	2	12	97
Stage-IV ($N_{sw} = 20$)	18	2	16	33	18	2	16	161
Stage-V ($N_{sw} = 24$)	22	2	20	41	22	2	20	221

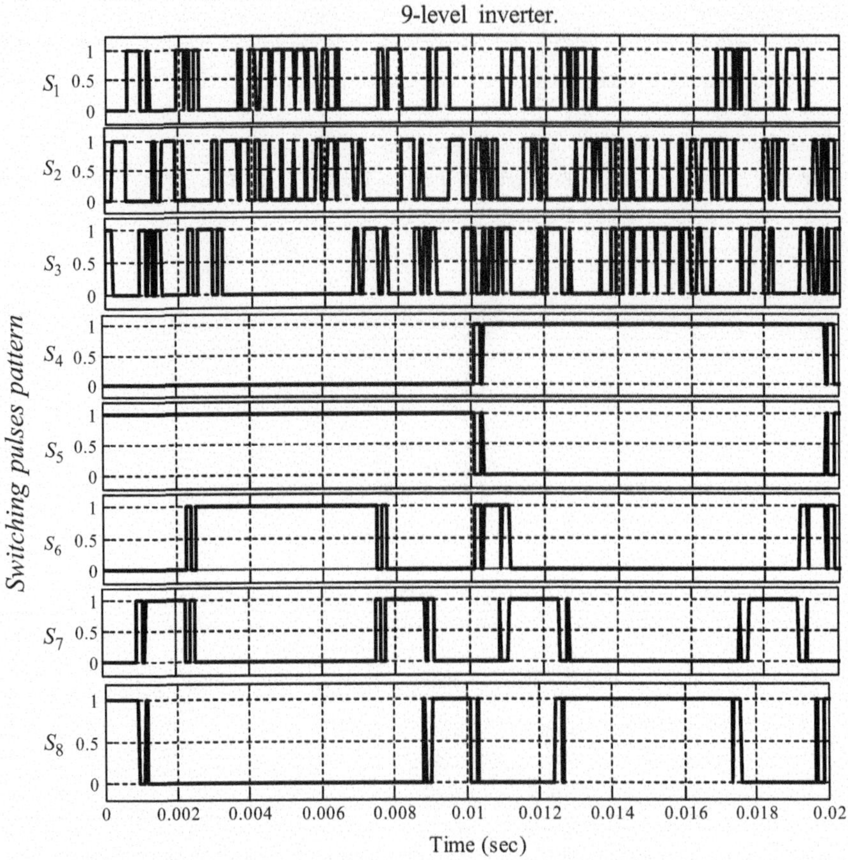

Fig. 6 Pulse pattern of each power switches in proposed 17-level inverter

5 Experimental and Simulation Results

The proposed 9-level symmetrical topology is simulated and executed using level-shifted PWM as control technique with considering the mentioned parameters such as magnitude of all the four DC sources as $V_{lu} = V_{ll} = V_{lu} = V_{ll} = 40$ V, $R = 50\ \Omega$, L = 100 mH at a 50 Hz fundamental frequency and operating at switching frequency of 3 kHz generates 160 V output stepped voltage. The simulated result has been incorporated in Fig. 8 where the output stepped voltage and %THD of stepped output voltage (phase voltage) is shown. The prototype model of the proposed multi-level inverter is established in the laboratory using dSPACE-1103 platform. Components employed for prototype model are as follows:

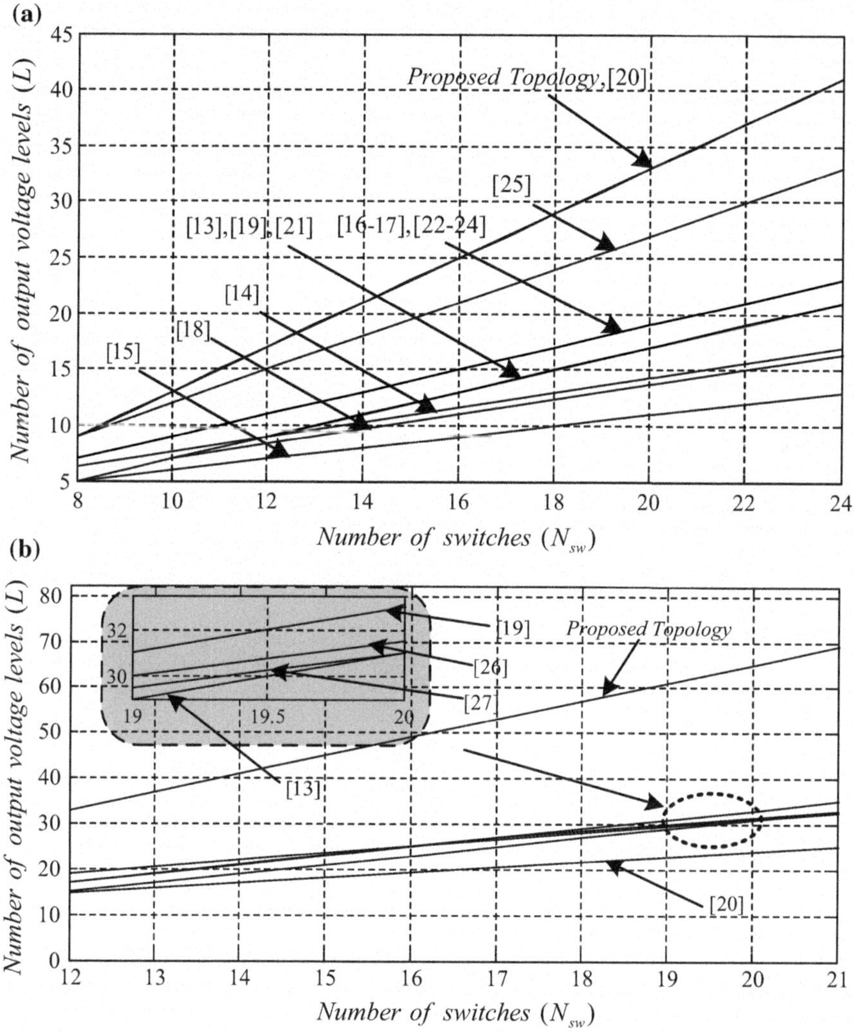

Fig. 7 **a** Graph indicating number of switches (N_{sw}) versus number of output voltage levels (L) with existed topologies for symmetrical configurations. **b** Graph indicating number of switches (N_{sw}) versus number of output voltage levels (L) with existed topologies for asymmetrical configurations

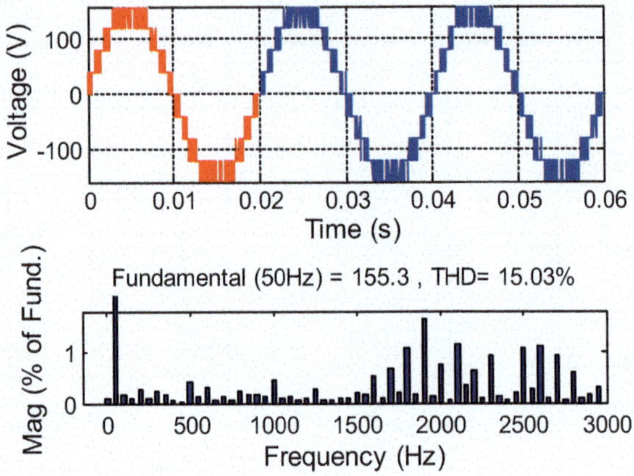

Fig. 8 Output voltage and %THD of output phase voltage of proposed 9-level symmetrical Inverter

(a) IGBT—Maker name—CT60AM.
(b) An opto-isolator TLP250 is used as driver.
(c) Isolated DC sources having magnitude as $V_{\mathrm{lu}} = V_{\mathrm{ll}} = V_{\mathrm{lu}} = V_{\mathrm{ll}} = 40$ V
(d) $R = 50\ \Omega$, $L = 100$ mH.

Figure 9a, b shows the experimental results of output voltage (160 V) of proposed symmetrical 9-level inverter for 3-cycles and 1-cycle respectively at $M_i = 1$ (Modulation Index).

6 Conclusions

The recent work proposed a 9-level single-phase inverter configuration employing the optimal number of components. The lesser number of power switches to generate the larger output voltage levels is described and compared with some recent reduced switch topologies. The proposed topology is simulated in MATLAB and the prototype of same is developed in the laboratory for symmetrical 9-level inverter. To verify the effectiveness of proposed MLI for both cases i.e. symmetrical and asymmetrical topologies, various other reduced switch MLI topologies are compared with the proposed topologies and found better in each corner regarding the total number of components, number of levels. Nevertheless, the generalized formulae for the proposed topology as well as the other compared reduced switch papers are incorporated as well.

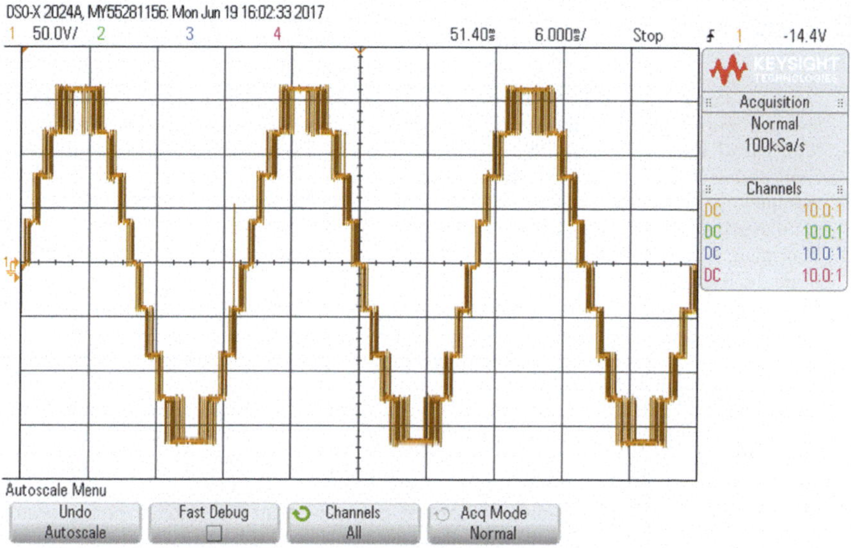

(a) Hardware result of voltage and current (3-cycles) of the

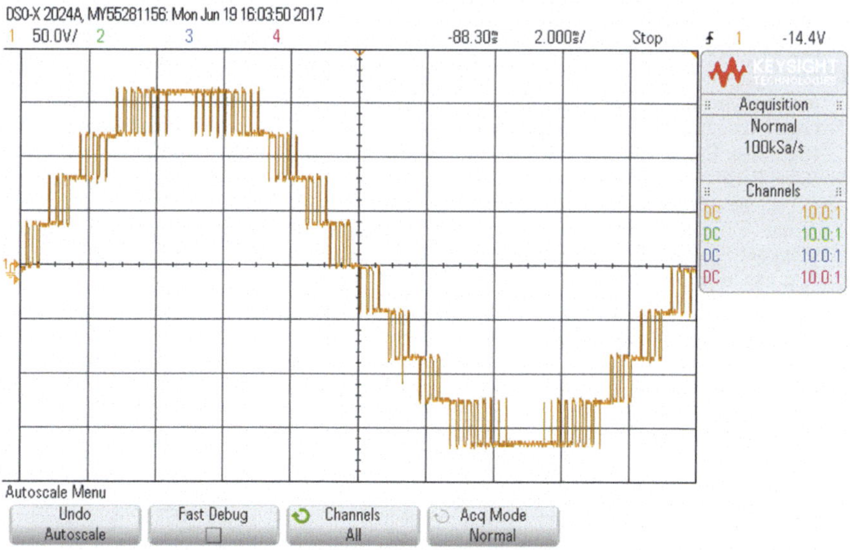

(b) Hardware result of voltage and current (1-cycle) of pro posed 9-level Inverter.

Fig. 9 Hardware results of 9-level proposed inverter

References

1. Liserre M, Sauter T, Hung JY (2010) Future energy systems: integrating renewable energy into the smart power grid through industrial electronics. IEEE Ind Electron Mag 4(1):18–37
2. Rodríguez J, Lai JS, Peng FZ (2002) Multilevel inverters: a survey of topologies, controls, and applications. IEEE Trans Ind Electron 49(4):724–738
3. Franquelo LG, Rodriguez J, Leon JI, Kouro S, Portillo R, Prats MAM (2008) The age of multilevel converters arrives. IEEE Ind Electron Mag 2(2):28–39
4. Rodriguez J, Bernet S, Steimer PK, Lizama IE (2010) A survey on neutral-point-clamped inverters. IEEE Trans Ind Electron 57(7):2219–2230
5. Malinowski M, Gopakumar K, Rodriguez J, Pérez MA (2010) A survey on cascaded multilevel inverters. IEEE Trans Ind Electron 57(7):2197–2206
6. Jing H, Corzine KA (2006) Extended operation of flying capacitor multilevel inverters. IEEE Trans Power Electron 21(1):140–147
7. Sotoodeh P, Miller RD (2014) Design and implementation of an 11-Level Inverter with FACTS capability for distributed energy systems. IEEE J Emerg Sel Top Power Electron 2(1):87–96
8. Babu NNVS, Fernandes BG (2014) Cascaded two-level inverter-based multilevel STATCOM for high-power applications. IEEE Trans Power Deliv 29(3):993–1001
9. Flores P, Dixon J, Member S, Ortúzar M, Carmi R, Barriuso P et al (2009) Static var compensator and active power filter with power injection capability, using 27-level. IEEE Trans Ind Electron 56(1):130–138
10. Khoucha F, Lagoun SM, Marouani K, Kheloui K, Benbouzid MEH (2010) Hybrid cascaded H-bridge multilevel-inverter induction-motor-drive direct torque control for automotive applications. IEEE Trans Ind Electron 57(3):892–899
11. Zheng Z, Wang K, Xu L, Li Y (2014) A hybrid cascaded multilevel converter for battery energy management applied in electric vehicles. IEEE Trans Power Electron 29(7):3537–3546
12. Gupta KK, Jain S (2014) Comprehensive review of a recently proposed multilevel inverter. IET Power Electron 7(3):467–479
13. Babei E, Hosseini SH (2009) New cascaded multilevel inverter topology with minimum number of switches. Energy Convers Manag 50:2761–2767
14. Ajami A, Jannati Oskuee MR, Toopchi Khosroshahi M, Mokhberdoran A (2014) Cascade-multi-cell multilevel converter with reduced number of switches. IET Power Electron 7(3):552–558
15. Gupta KK, Jain S (2012) Topology for multilevel inverters to attain maximum number of levels from given DC sources. IET Power Electron 5(4):435–446
16. Banaei MR, Jannati Oskuee MR, Khounjahan H (2014) Reconfiguration of semi-cascaded multilevel inverter to improve systems performance parameters. IET Power Electron 7(5):1106–1112
17. Banaei MR, Salary E (2011) Verification of new family for cascade multilevel inverters with reduction of components. J Electr Eng Technol 6(2):245–254
18. Thamizharasan S, Jeevananthan S, Baskaran J, Ramkumar S (2014) Cross-switched multilevel inverter using auxiliary reverse-connected voltage sources. IET Power Electron 7(6):1519–1526
19. Ajami A, Jannati Oskuee MR, Mokhberdoran A, Van den Bossche A (2014) Developed cascaded multilevel inverter topology to minimise the number of circuit devices and voltage stresses of switches. IET Power Electron 7(2):459–466
20. Gautam SP, Kumar L, Gupta S (2015) Hybrid topology of symmetrical multilevel inverter using less number of devices. IET Power Electron 8(11):2125–2135
21. Najafi E, Halim A, Yatim M, Member S (2012) Design and Implementation of a new multilevel inverter topology. IEEE Trans Ind Electron 59(11):4148–4154
22. Gupta KK, Jain S (2014) A novel multilevel inverter based on switched dc sources. IEEE Trans Ind Electron 61(7):3269–3278
23. Jain S, Gupta KK (2013) Multilevel inverter topology based on series connected switched sources. IET Power Electron 6(1):164–174

24. Babaei E, Farhadi Kangarlu M (2013) Cross-switched multilevel inverter: an innovative topology. IET Power Electron 6(4):642–651
25. Salary E, Jannati Oskuee MR, Najafi-Ravadanegh S (2015) Creative design of symmetric multilevel converter to enhance the circuit's performance. IET Power Electron 8(1):96–102
26. Hinago Y, Koizumi H (2010) A single-phase multilevel inverter using switched series/parallel DC voltage sources. IEEE Trans Ind Electron 57(8):2643–2650
27. Samadaei E, Gholamian SA, Sheikholeslami A, Adabi J (2016) An envelope type (E-Type) module: asymmetric multilevel inverters with reduced components. IEEE Trans Ind Electron 63(11):7148–7156
28. Jana K, Biswas S, Kar Chowdhury S (2016) Dual reference phase shifted PWM technique for a N-level inverter based grid connected solar photovoltaic system. IET Renew Power Gener 10(7):928–935
29. Mahato B, Raushan R, Jana KC (2017) Modulation and control of multilevel inverter for an open-end winding induction motor with constant voltage levels and harmonics. IET Power Electron 10(1):71–79
30. Jana KC, Biswas SK (2015) Generalised switching scheme for a space vector pulse-width modulation-based N-level inverter with reduced switching frequency and harmonics. IET Power Electron 8(12):2377–2385

Bidyut Mahato (Senior Research Fellow) pursuing Ph.D. from **IIT(ISM)**, Dhanbad, received B.Tech and M.E degrees from **GNIT**, **Kolkata**, West Bengal and **BIT**, **Mesra**, Ranchi in the year 2011 and 2014 respectively. He is Member of **International Association of Engineers, Associate Member** of **IETE, the IRED** and **The Institution of Engineers, Life Member** of **SSI** and **Life Member** of **ISTE**. His research interest includes Power electronic converters, multilevel inverter, Power factor correction and hybrid electric vehicles.

Sudhanshu Mittal received his **B.Tech** degree from **Sam Higginbottom Institute of Agriculture, Technology and Sciences** in Department of Electrical Engineering **2013**. Currently, he is pursuing his M.Tech degree from **IIT(ISM), Dhanbad** in Dept. of Electrical Engineering. His area of interest includes converters, inverters and Multilevel Inverters.

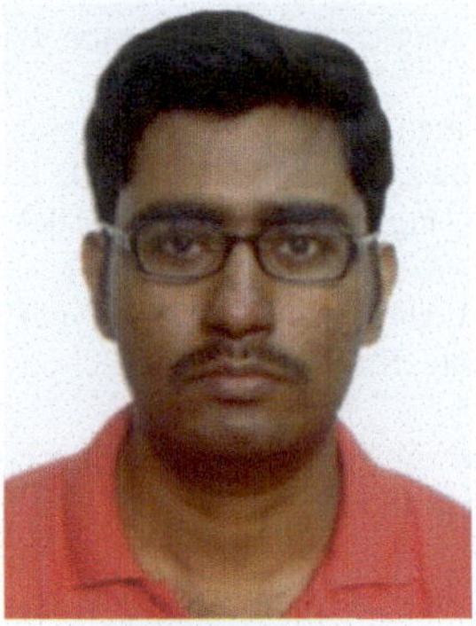

Saikat Majumdar (Junior Research Fellow) received his B.Tech and M.Tech from **AOT** and **IIT(ISM)**, Dhanbad in the year 2014 and 2016 respectively. He is working towards his Ph.D. from 2016 in the area of multilevel inverter, renewable energy sources and drives.

Kartick Chandra Jana (Engg, Dr.) received his M.Tech and Ph.D. degree from **NIT Durgapur** and **Jadavpur** University in 2003 and 2013 respectively. He is currently serving as Assistant Professor in **IIT(ISM)**, Dhanbad from June 2012. He has been working in **Birla Institute of Technology, Mesra** from July 2003 till May 2012. His research is in area of Modelling and Design of Multilevel Inverter, Real-time control of power electronics devices, Design and Implementation of efficient power.

Paresh Kumar Nayak received the **Ph.D**. degree in Electrical Engineering from the **IIT Kharagpur**, India, in **2014**. He received the **B.E**. degree in electrical engineering from **University College of Engineering, Burla**, Sambalpur, in **2000** and the **M.Sc**. degree from the Department of High Voltage of Engineering, **IISc, Bangalore**, India, in **2003**. Currently, he is an Assistant Professor in the Department of Electrical Engineering, **IIT(ISM)**, Dhanbad, India. His current research interest is power system relaying, analysis of high-frequency behavior of transformer winding.

Enhancement of Fatigue Life of TIG-Welded Joint by Friction Stir Processing

Ajaya Bharti and Hariom Tripathi

Abstract In the present work, Aluminium alloy sheets graded as Al6061 having 6 mm thickness were TIG welded as a butt joint. In order to modify the microstructure, the joint of welded samples were subjected to friction stir processing (FSP). The effect of FSP on the microstructure in the welded and heat affected region (HAZ) was characterized by optical microscopy. Further, the mechanical strength, hardness and fatigue life was determined through mechanical test, hardness test and fatigue test. The fatigue life tests were performed at constant amplitude loading by taking stress ratio equals to zero. It has been found that the FSP increases the tensile strength approximately 5–13% whereas hardness 5–10%. Similarly, there has been found approximately 30–60% improvement in fatigue strength. It is attributed that such kind of changes were found due to grain refinement in the welded region and its vicinity. Other reasons might be due to the modification in geometry of weld toe. It has been observed that the weld defects like porosity, hot cracking and lack of wetting have been also reduced.

Keywords Al6061 · TIG · Friction stir processing · Heat affected zone
Fatigue strength

1 Introduction

The Aluminium and its alloys are extensively used in automotive parts, structural components, aerospace parts and railways [1], because they have good corrosion resistance, high specific strength, high toughness and better formability. Though, the Aluminium and its alloys are generally difficult to weld. Therefore, TIG welding is largely used in joining Aluminium and its alloys. The welded samples generally have some drawbacks [2, 3] like the presence of porosity, distortion, lack of wetting, residual stresses and hot cracking due to which their welding strength decreases,

A. Bharti (✉) · H. Tripathi
Department of Applied Mechanics, MNNIT, Allahabad 211004, India
e-mail: abharti@mnnit.ac.in

J. Chattopadhyay et al. (eds.), *Renewable Energy and its Innovative Technologies*,
https://doi.org/10.1007/978-981-13-2116-0_5

especially on the application of cyclic load. In the butt welded joint, toe of welds is the critical zone for fatigue strength, due to geometrical discontinuity and large stress concentration. There are several techniques by which one can improve the fatigue behaviour of TIG-welded samples, such as re-melting, hammering and shot peening [4, 5].

It has been observed that the mechanical strength of metals, alloys and composites can be enhanced by the refinement of grain of metals and alloys. Further, the refinement of grain can be done by the Severe Plastic Deformation (SPD) techniques [6–10]. The Friction Stir Process (FSP) is one of the most important SPD techniques which are used for producing ultrafine grain materials. The FSP appears as a substitute for traditional methods (hammering, re-melting) for enhancing mechanical strength of the welded specimen. The FSP modifies the existing and localized microstructure and hence produces the better mechanical properties. Recently, FSP emerged as an alternative process to improve fatigue behaviour of TIG-welded region. The Friction stir processing (FSP) [11–14] is used for micro-structural modification and enhancing mechanical strength, which is based on the principal of friction stir welding (FSW). In 1991, the FSW technique was first applied by Welding Institute Cambridge (UK) [15–18]. It is a solid state welding technique used for joining soft materials like Aluminium and its alloys. The optimization of the FSP parameters [19, 20] and design of tool can control the microstructure within the processed region and thus the desired mechanical properties.

There are very limited literatures available and have been focused on the effect of FSP on the mechanical and microscopic behaviour of TIG-welded Aluminium alloys. Therefore, the objective of present research work is to determine the fatigue behaviour of TIG butt welded Al6061 alloy before and after the FSP. Additionally, tensile strength, hardness profile, microstructural properties and XRD analysis were also identified to verify the fatigue life behaviour.

2 Experiments

2.1 TIG Welding and FS Processing

The alloy Al6061 rolled plates having size $6 \times 200 \times 90$ mm were used for experimental work. The plates of the alloy were solution treated at temperature 520 °C for 2 h to nullify the previous history of material and to have single phase in the material. The detail of the chemical composition of the Al6061 Aluminium was found as shown in Table 1.

The solution-treated plates were TIG-welded using welding machine (make: TIG ARC300) as a single 'V' butt joint as shown in Fig. 1. The filler material used was ER4043, having 2 mm diameter wire, and welds were shielded by pure argon. The welding parameters and chemical composition of ER4043 were taken as per the reference [21]. The TIG welds were produced normally in the direction of rolling

Table 1 Chemical composition Al6061 Aluminium

Elements	Mg	Si	Ti	Mn
Weight (%)	1.2	0.6	0.04	0.07
	Zn	Fe	Cr	Cu
	0.09	0.4	0.2	0.2
	Rest is Aluminium			

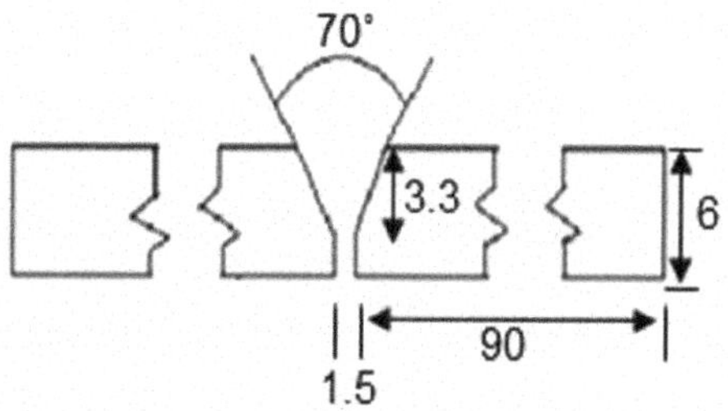

Fig. 1 Geometry and dimension of joint (dimensions in mm)

in a single pass, joining two plates of 200 × 90 × 6 mm, with width of 90 mm on each side, as illustrated in Fig. 1. The root of first weld was cleaned with the help of grinding machine before second weld bead deposited on back side of first weld to obtain a completely penetrated butt weld. TIG-welded region in small slice 20 × 180 mm of Al6061 Aluminium alloy strip is as shown in Fig. 2.

After the TIG welding of Aluminium alloy Al6061 plates, friction stir post-processing of TIG-welded plates were performed on vertical milling machine, by holding the TIG-welded plates on the milling machine bed. The friction stir processing were performed both sides of welded plates. There was some burrs formation on the processed zone which was removed by the help of machining. The FSP tool, as shown in Fig. 3, was rotated at the speed of 635 rpm and 120 mm/min using vertical milling machine to make sure 3 mm of effective penetration of pin inside the welded zone.

2.2 *Microstructure Characterization and Testing*

In order to investigate the change in the microstructure or morphology due to welding and friction stir processing, the samples were extracted from the different regions like base material (BM), heat affected zone (HAZ), welding region (WR) of the welded and FSP-processed samples. These samples were prepared for optical microscopy as per the ASTM E3 [22].

To investigate the change in the Micro-hardness, Tensile strength and Fatigue life behaviour samples were prepared according to ASTM 384 [23], ASTM E8 [24] and ASTM E466 [25]. The micro-hardness test was carried out according to on the polished sample surface by the help of Vickers micro-hardness tester. For drawing

(a)

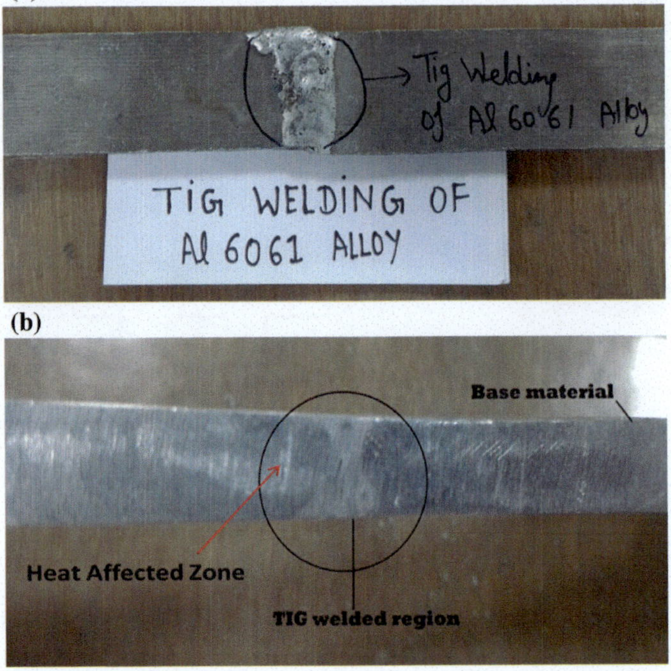

Fig. 2 **a** TIG-welded region in Al6061 alloy plates **b** TIG-welded region after grinding and polishing

Fig. 3 Geometry of FSP Tool and real tool

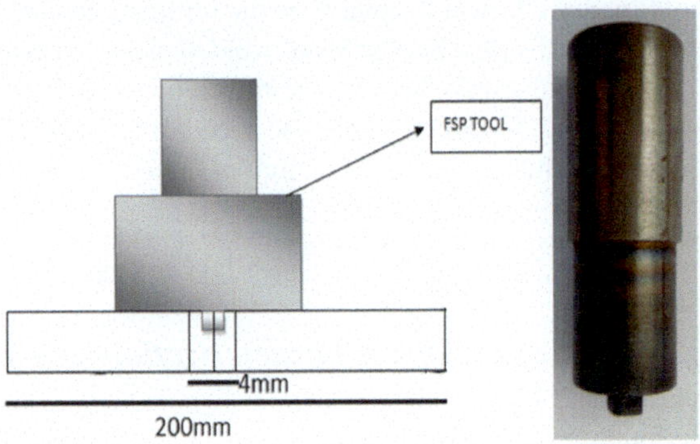

hardness profile indents were taken by the application of 0.2 kg load for 10 s dwell period. To plot the hardness curve minimum of 25 indents were taken.

Tensile testing was carried out on a servo-hydraulic UTM machine (make: BISS, India) having 25 kN load capacity using 1 mm/min as a testing speed Fatigue test was also performed on a servo-hydraulic UTM machine. The Aluminium alloy solution treated, welded plates, welded and FSP processed were extracted into the small slices of 20 × 200 mm, transversely for tensile and fatigue life testing. The Fatigue life tests were performed by applying the Sinusoidal load at frequency 20–25 Hz, stress ratio (R) equal to zero and stress range 70–260 MPa. The results of fatigue test were plotted on stress range versus the number of failure cycles (S-N) curves, all test were performed at room condition.

3 Results and Discussion

3.1 Microstructure

In order to make morphological analysis of TIG-welded samples before and after friction stir processing, micrographs obtained by optical microscope are illustrated in Fig. 4a–e. The BM consists of elongated grains in the direction of rolling, as illustrated in Fig. 4a, with an average size of the grain in the range of the 85–120 μm in the direction of rolling, measured with the help of Heyn-linear intercept method [26]. In HAZ, there was slightly distorted microstructure obtained as illustrated in Fig. 4b. The weld region is formed by approximately equiaxed grain as illustrated in Fig. 4c, with an average size of the grain in the range of the 100–130 μm, due to uniform cooling at the centre of the layers. Figure 4d–e illustrates the microstructure observed after TIG welding followed by friction stir processing. There was a large grain refinement observed in the HAZ and welded zone after FSP as shown in Fig. 4d and e, respectively. The average size of grain in the range of 4.5–10 μm and 11.5–16.9 μm were observed within the heat affected zone and welded region (WR) respectively, which was measured by the intercept method as mentioned earlier.

3.2 Hardness

The micro-hardness test results of welded and welded followed by FSP regions are shown as Fig. 5. The indentations were made in the three regions, i.e. welded, HAZ and base materials. This image shows that there is a significant reduction in Vickers hardness value in the HAZ and welded pool region, because coarsening of grains and dissolution of strengthening precipitates within the welded and heat affected zone take place. However, there are some hardness peaks obtained in the welded zone due to some particle precipitation during the cooling. Before processing the hardness test values of the base material (16–17 mm from the weld bead centre) obtained in between the range of 101–112 HV and for heat affected zone and welded pool region it was obtained in between the range of 65–84 HV. After the welding and FSP, there is approximately 5–10% increase in hardness values obtained within the HAZ (in between 6 and 16 mm from the weld bead centre) and weld pool region (−5 to 5 mm).

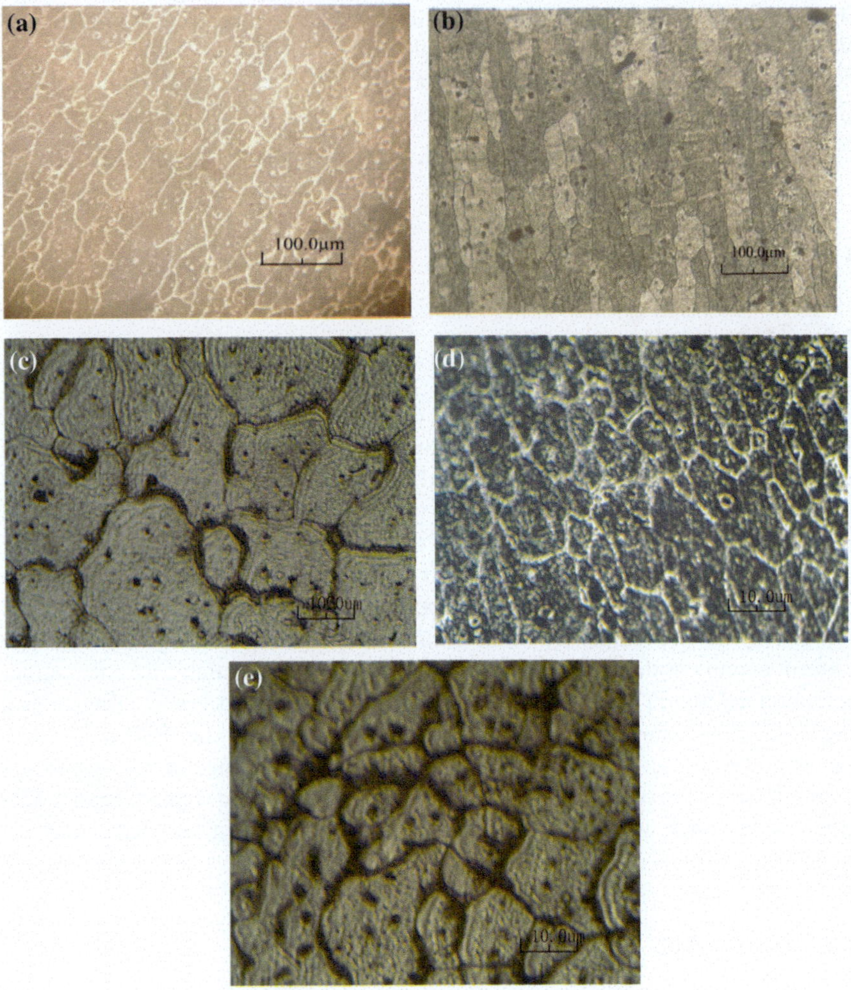

Fig. 4 Optical microscope images of TIG welds at 200× **a** base material plate (BM), **b** heat affected region (HAZ), **c** welded region (WR) **d** HAZ + FSP **e** WR + FSP

The regions with an increase in hardness in FS processed zones are attributed to the grain refinement or plastic deformation induced by the FSP.

3.3 Tensile Strength

The tensile strength tests results of welded samples and Welded followed by friction stir processing are shown in Table 2. This table also shows the welding efficiency [27].

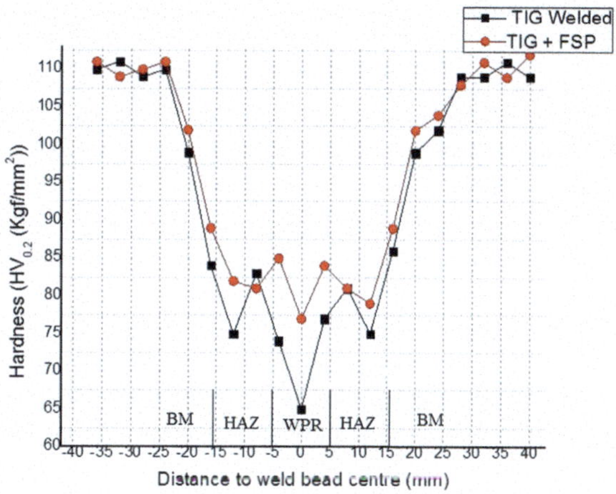

Fig. 5 Hardness profile of TIG welded and TIG+FSP: BM-base material; HAZ—heat affected zone; WPR—weld pool region

Table 2 Tensile properties of ST, ST+TIG, ST+TIG+FSP

Specimen type	Yield strength (MPa)	Ultimate strength (MPa)	Elongation (%)	Weld efficiency
ST	252	311	19	–
TIG	157	202	8	64.95
TIG+FSP	163	217	13	69.77

The tensile strength of solution-treated material showed yield strength of 252 MPa and ultimate strength of 311 MPa. On the other hand, TIG-welded plates showed a yield tensile strength and ultimate tensile strength of 157 and 202 MPa respectively. It means weld efficiency is obtained less than one. There is approximately 36–40% reduction in tensile strength for TIG-welded joints as compared to solution-treated samples. This may be due to the presence of welding defects and reduction in strain hardening.

The welded followed by FSP samples shows a small enhancement in tensile strength and strain than that of the TIG-welded sample. However, it has less strength and strain than that of the solution-treated material. The welded followed by FSP sample showed a yield tensile strength and ultimate tensile strength of 163 and 217 MPa respectively. This increase in tensile strength is due to the removal of the welding defects and the grain refinement.

3.4 *Fatigue Properties*

Similar to tensile tests, the fatigue tests results of base material, i.e. solution-treated (ST), TIG-welded and TIG followed by friction stir processed samples was subjected to cyclic loading. For each kind of samples, the results were plotted between

Nominal stress range ($\Delta\sigma_{nom}$) and the number of failure cycles (N_f). This curve is known as Wohler curve (S-N curve). It is observed from the S-N curve that there is approximately 40–60% decrease in fatigue strength of TIG-welded samples relative to base material samples in between 10^4 and 10^6 numbers of cycles; this is due to weld defects and stress concentration at the weld toe. From the Wohler curve it is observed that after FSP there is a significant increase in fatigue resistance of TIG-welded samples. The FSP-welded samples shows approximately 30–60% increase in fatigue strength as compared to TIG-welded samples. However, fatigue strength of TIG + FSP samples remains lower relative to base material samples.

The main causes of enhancement in fatigue life by friction stir processing are the modification in geometry as a result of the increase in the radius of curvature, lack of wetting and modification in the microstructure. The FS Processing mainly retards the crack initiation step due to micro-structural modification.

4 Conclusion

In this present work, it is concluded that the TIG welding reduces the tensile strength as well as fatigue strength of the welded plate as compared to base material. However, with the FS processing of welded region, the tensile strength as well as fatigue strength can be improved due to reduction of the grain size and removal of defects caused due to welding.

References

1. Maisonnette D, Suery M, Neliasa D, Chaudet P, Epicier T (2011) Effects of heat treatments on the microstructure and mechanical properties of a 6061 aluminum alloy. Mater Sci Eng, A 528:2718–2724
2. Wang X, Wang K, Shen Y, Hu K (2008) Comparison of fatigue property between friction stir and TIG welds. J Univ Sci Technol Beijing 15:280
3. Kah P, Rajan R, Martikainen J, Suoranta R (2015) Investigation of weld defects in friction-stir welding and fusion welding of aluminum alloys. Int J Mech Mater Eng 10:26
4. Haagensen PJ (2011) Fatigue strength improvement methods. Fract Fatigue Welded Joints Struct 11:297–329
5. Hatamleh O (2008) The effects of laser peening and shot peening on mechanical properties in friction stir welded 7075-T7351 aluminum. J Mater Eng Perform 17:688–694
6. Ding R, Chung C, Chiu Y, Lyon P (2010) Effect of ECAP on microstructure and mechanical properties of ZE41 magnesium alloy. Mater Sci Eng, A 527:3777–3784
7. Kumar S, Raghu T (2013) Mechanical behavior and microstructural evolution of constrained groove pressed nickel sheets. J Mater Process Technol 213:214–220
8. Saito Y, Tsuji N, Utsunomiya H, Sakai T, Hong R (1998) Ultra-fine grained bulk aluminum produced by accumulative roll-bonding (ARB) process. Scripta Mater 39:1221–1227
9. Kamachi M, Furukawa M, Horita Z, Langdon TG (2003) Equal-channel angular pressing using plate samples. Mater Sci Eng, A 361(1–2):258–266

10. Ardakani M, Amirkhanlou S, Khorsand S (2014) Cross accumulative roll bonding—A novel mechanical technique for significant improvement of stir-cast Al/Al_2O_3 nanocomposite properties. Mater Sci Eng A591:144–149
11. Santelia ML, Engstrom T, Storjohann D, Pan TY (2005) Effects of friction stir processing on mechanical properties of the cast aluminum alloys A319 and A356. Scripta Mater 53:201–206
12. Ohannes LB, Mishra RS (2007) Multiple passes of friction stir processing for the creation of superplastic 7075 aluminum. Mater Sci Eng, A 464:255–260
13. Cavaliere P, De Marco PP (2007) Friction stir processing of AM60B magnesium alloy sheets. Mater Sci Eng, A 462: 393–397
14. Dieguez T, Burgueño A, Svoboda H (2012) Superplasticity of a friction stir processed 7075-T651. Aluminum Alloy. Procedia Mater Sci 1:110–117
15. Mishra RS, Ma ZY (2005) Friction stir welding and processing. Mater Sci Eng 50:1–78
16. Commin L, Dumont M, Masse J, Barrallier L (2009) Friction stir welding of AZ31 magnesium alloy rolled sheets. Acta Mater 57:326–334
17. Liu G, Murr LE, Niou CS, McClure JC, Vega FR (1997) Microstructural aspects of the friction-stir welding of 6061-T6 aluminum. Scripta Mater 37:355–361
18. Thomas WM, Nicholas ED, Needham JC, Murch MG, Dawes CJ (1991) International Patent Application PCT/GB92/02203 and GB Patent Application No. 9125978. 8 December 1991
19. Svensson LE, Karlsson L, Larsson H, Karlsson B, Fazzini M, Karlsson J (2000) Microstructure and mechanical properties of friction stir welded aluminum alloys with special reference to AA5083 and AA6082. Sci Technol Weld Joining 5:285–297
20. Jata KV, Sankaran KK, Ruschau JJ (2000) Friction-stir welding effects on microstructure and fatigue of aluminum alloy 7050-T7451. Metallurgical Mater Trans A 31:2181–2192
21. Ambriz RR, Barrera G, García R, López VH (2010) The microstructure and mechanical strength of Al-6061-T6 GMA welds Obtained with the modified indirect electric arc joint. Mater Des 31:2978–2986
22. ASTM Standard E3-11, Standard Guide for Preparation of Metallographic Specimens
23. ASTM Standard E384, 2011e1, Standard Test Method for Knoop and Vickers Hardness of Materials, ASTM International
24. ASTM Standard E8/E8M-16a, Standard Test Methods for Tension Testing of Metallic Materials
25. ASTM Standard E466-15, Standard Practice for Conducting Force Controlled Constant Amplitude Axial Fatigue Tests of Metallic Materials
26. ASTM Standard E112, Standard Test Methods for Determining Average Grain Size
27. Elangovan K, Balasubramanian V (2008) Influences of post-weld heat treatment on tensile properties of friction-stir-welded AA6061 aluminum alloy joints. Mater Charact 59:1168–1177

Dr. Ajaya Bharti is working as Associate Professor since last 13 years with the institute named as "Motilal Nehru National Institute of Technology Allahabad", Allahabad, India-211004 having Ph.D., M.Tech. (material science and engineering), B.Tech. (mechanical) qualifications, and teaching UG and PG students as well as doing research in the areas of material science, fatigue and fracture mechanics, wear, corrosion, severe plastic deformation (SPD), physical metallurgy, biomaterials, structural health monitoring, synthesis and characterization of advanced materials.

Mr. Hariom Tripathi is associated as a Ph.D. reserch scholar with the Institute named as "Motilal Nehru National Institute of Technology Allahabad", Allahabad, India-211004 under guidance of Dr. Ajaya Bharti. He is doing his research in the area of "Develoment and characterization of nanostructured alloys". He is having qualification B.Tech. (mechanical), M.Tech. (materials science and engineering).

Exergy Analysis and Irreversibility of Combustion Process of an Auxiliary Boiler for Marine Application

Jitendra Singh Pal, S. N. Sapali and T. R. Anil

Abstract Auxiliary boiler is used on board vessels for the purpose of steam generation. In marine vessels where boiler is running on liquid fuel for the combustion purpose, the irreversibility leads to exergy loss and high operating cost. This exergy analysis of auxiliary boiler (10.5 bar capacity) constitutes a good working tool to access the thermodynamic quality of a system. This study reveals the thermodynamic condition of the system by following a chain of events like irreversibility, degradation of the energy, and loss occur. Numerical analysis of burner for the mass fraction and temperature gradient is carried out for viscous dissipation, heat conduction, mass diffusion, and chemical reaction. It suggests 15% better optimization of the systems and rational operation of the installation than the existing system by exergy addition, exergy losses, and exergy consumption on the measured operating data and the thermodynamic state of the plant components. It reveals that the rate of exergy destruction by chemical reaction can be reduced when the flame temperature is high. This analysis of the auxiliary boiler will contribute to an exergy analysis of combustion process, identification of irreversibility leading to highly energy efficient boiler operation and design modification.

Keywords Exergy · Irreversibility · Optimizer · Preheating of air · Mass fraction

J. S. Pal (✉)
Marine Engineering Department, Tolani Maritime Institute, Pune, India
e-mail: Jitendrasinghpal2011@gmail.com

S. N. Sapali
Mechanical Engineering Department, College of Engineering, Pune, India
e-mail: sns.mech@coep.ac.in

T. R. Anil
Gogte Institute of Technology, Belagavi, India
e-mail: tranil@git.edu

J. Chattopadhyay et al. (eds.), *Renewable Energy and its Innovative Technologies*,
https://doi.org/10.1007/978-981-13-2116-0_6

1 Introduction

The purpose of marine boiler is to produce steam to drive turbo-generator and cargo pumps, heating of fuel and lubricating oil for purification purpose, cabin heating, tank heating, fire-fighting, and other service needs. A marine steam boiler is highly reliable in operation since sea-going vessels usually make long voyages without stopping at a port. In this study the Model RX 4 of Thermax make installed in ship-in-campus Training Ship Prabhu-Vidhya is used for the energy and exergy analysis. It is a once through, water tube boiler with reverse flame technology in which the water passes along the tubes only once and gets completely evaporated. The Furnace section of the boiler contains a membrane coil where fuel is burnt. This coil is housed in a jacket assembly consisting of inner jacket and outer jacket. Pressure jet burner is mounted on the top cover of the outer jacket. Apart from the main boiler section, economizer, and heat optimizer are provided before the feed water pump. Economizer is a shell and tube heat exchanger and optimizer is tube in tube-type heat exchanger. The necessary fuel handling system is also provided along with the boiler. Fuel is burnt is the membrane coil by the pressure jet burner using the air from combustion air fan. At the outlet of the boiler, the flue gas is exhausted through the economizer to recover further heat from the flue gas. Water from the feed water tank enters the economizer through the optimizer before entering the feed water pump. The economizer outlet is again connected to the optimizer. By this water circuit, the temperature of water entering the feed water pump is maintained below 80 °C which is the maximum temperature the reciprocating pump can withstand. This boiler is working on Light Diesel Oil (LDO), which is supplied by a gear pump having an output capacity of 2000 L/h against 50 psi delivery pressure. One mesh type filter of 80 × 80 mesh is provided just at the outlet of the main storage tank.

2 Literature Review

Moran [1] gives the definition and second law efficiencies for efficient energy use. As per him the availability is the maximum work that can be extracted from the combined system of control mass and environment as the control mass passes from a given state to the dead state. Availability, symbolized by A, is an extensive property calculated by

$$A = (E - U_0) + P_0(V - V_0) - T_0(S - S_0), \tag{1}$$

where E (=U + KE + PE), V, and S denote, respectively, the energy, volume, and entropy of the control mass at the given state and U_0, V_0, and S_0 are the same properties when the control mass is at rest at the dead state.

The second law efficiency € is a ratio of Availability out in product to Availability in [2]. Another class of second law efficiencies is composed of task efficiencies. Task efficiency is the ratio of the theoretical minimum input required by the first and second laws to accomplish some task to the actual input for a particular means.

Kotas [3] explained that the irreversibility inherent in the adiabatic combustion process corresponds to a large proportion of the original exergy of the fuel. To reduce this irreversibility a reduction in the rate of increase in entropy is required, which is always associated with an increase in the maximum temperature of the products and it can be achieved by isochoric combustion, oxygen enrichment and preheating of the reactants.

The chemical exergy of liquid fuels:

$$\dot{\varphi} = 1.0401 + 0.1728\, h/c + 0.0432\, o/c + 0.2169\, s/c\, (1 - 2.0628\, h/c) \tag{2}$$

3 Exergy Analysis

3.1 Chemical Exergy

Refer Fig. 1, this RXA-04 of Thermax, three pass, reverse flue, coil-type steam generator installed in training ship Prabhu-Vidya is studied for the second law analysis and identification of irreversibility present with numerical analysis of burner.

The fuel used is light diesel oil of the following chemical composition by mass: $c = 0.871$, $h = 0.133$, $s = 0008$, $o = 0$ and the net calorific value of the fuel is = 42800 kJ/kg. The chemical exergy of this fuel is calculated by the Eq. (2) [4]:

$$\dot{\varphi} = 1.0401 + 0.1728\,(0.133/0.871) + 0.2169\,(0.008/0.871)\,\{1 - 2.0628\,(0.133/0.871)\}$$

$$\dot{\varphi} = 1.06785 \text{ kJ/kg}$$

Fig. 1 Auxiliary boiler

As this light diesel oil contains no moisture and the effect of sulfur is included in correlation:

$$'E^0 = \phi\,(\mathrm{NCV})^0 \tag{3}$$

$'E^0 = 45{,}704$ kJ/kg is the chemical exergy of the light diesel oil used for combustion purpose.

3.2 Air Heating Increases Available Exergy

Cold air from the combustion air fan is passed through the annulus space between inner and outer jacket to pre-heat and in recover heat from the flue gas. At the outlet of the boiler, the flue gas is exhausted though the economizer. For the constant pressure process and unit mass of the air being compressed by air fan is as follows [5]:

$$W_{\max} = C_p\left[(T_2 - T_1) - T_0 \log T_2/T_1\right] \tag{4}$$

The parameter of the auxiliary boiler state that we have $T_0 = 295$ K, $T_1 = 308$ K, $T_2 = 458$ K, $C_p = 1005$ J/kg °C. Thus $W_{\max} = 3.11 \times 10^4$ J/kg.

The total heat added is $C_p\,[(T_2 - T_1)] = 15 \times 10^4$ J/kg. So the increase in unavailable energy is $(15 - 3.311) \times 10^4$ J/kg $= 11.689 \times 10^4$ J/kg. It shows that the heat-addition process causes an increase in both the available and unavailable energy of the air.

3.3 The Second-Law Efficiency and Irreversibility of Auxiliary Boiler

In this RX 4 Auxiliary boiler, heat is transferred from the products of light diesel oil combustion to the steam. The temperature of the products of combustion decreases from 900 to 450 °C, while the pressure remains constant at 0.1 MPa. The average constant pressure specific heat of the product of combustion is 1.09 kJ/kg K. The water enters into the water tubes at 130 °C and leaves at 200 °C.

The ratio of the mass flow of products to the mass flow of water

$$M_{\mathrm{prod}}/M_{\mathrm{H_2O}} = (h_e - h_i)_{\mathrm{H_2O}}/(h_i - h_e)_{\mathrm{prod}}$$

$$M_{\mathrm{prod}}/M_{\mathrm{H_2O}} = 4.67474 \tag{5}$$

The increase in availability of the water is, per kilogram of water,

$$\dot{\phi}_2 - \dot{\phi}_1 = (h_2 - h_1) - T_0\,(S_2 - S_1)$$

$$\dot{\phi}_2 - \dot{\phi}_1 = 748.09\ \text{kJ/kg H}_2\text{O} \tag{6}$$

The decrease in availability of the products, per kilogram of water is

$$M_{prod}/M_{H_2O}\ (\dot{\phi}_3 - \dot{\phi}_4) = M_{prod}/M_{H_2O}\ [(h_3 - h_4) - T_0\,(S_3 - S_4)] \tag{7}$$

$$M_{prod}/M_{H_2O}\ (\dot{\phi}_3 - \dot{\phi}_4) = 1618.6089\ \text{kJ/kg H}_2\text{O} \tag{8}$$

Therefore, the second-law efficiency is,

$$N_{2\text{nd law}} = 748.09/1618.6089$$
$$N_{2\text{nd law}} = 0.3740$$

The process irreversibility per kilogram of water is,

$$I/M_{H_2O} = (\dot{\phi}_1 - \dot{\phi}_2) + M_{prod}/M_{H_2O}\ (\dot{\phi}_3 - \dot{\phi}_4)$$

$$I/M_{H_2O} = 870.5189\ \text{kJ/kg H}_2\text{O} \tag{9}$$

4 Numerical Analysis of Burner

A Pressure jet burner is mounted on the top cover of the furnace section of this RX 4 model auxiliary boiler. Mesh generation on the burner and furnace section with the hexahedrons wall-adapted is carried out with Ansys18.1 code. The numerical model of the burner is studied by two-step chemical reaction and the property value based on temperature gradients, mass fraction and species concentration.

The kinetic parameters are illustrated in Fig. 2, which reveals that the CO_2 concentration is more on the tip of the nozzles. As the temperature increases in the furnace the good oxidation takes place. It is understood that the low concentration species are not responsible for the entropy generation. The structure of a laminar non-premixed flame predicted which shows that the highest temperature zone is located in the burner port area [6].

Figure 3 illustrates the temperature contour at the tube outer walls which shows the temperature variation. The fuel droplets from the pressure jet burner mix with the hot air within the combustion chamber. The drops are heated up as results of convective heat transfer and temperature radiation of the hot chamber walls.

Figure 4 indicates that the C_8H_{18} mass fraction is distributed evenly in the whole combustion chamber, due to good atomization and mixing with air.

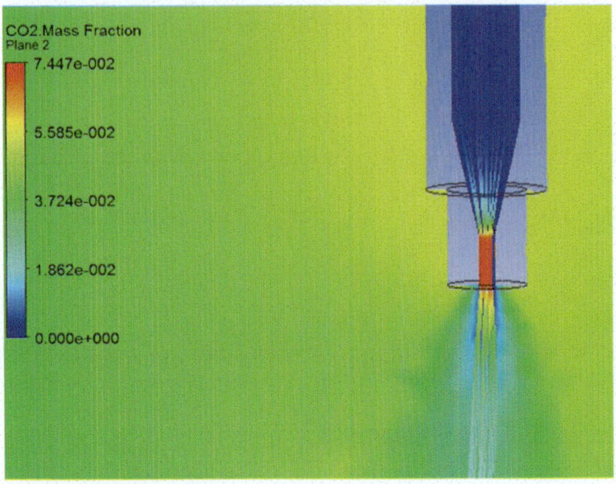

Fig. 2 CO_2 Mass fraction distribution

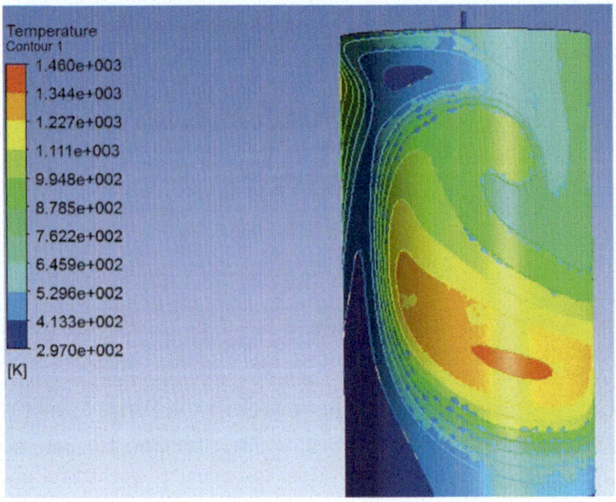

Fig. 3 Temperature contour of tube walls on the auxiliary boiler

5 Conclusions

Exergy analysis of the auxiliary boiler reveals that the second law efficiency is increased by heating the combustion air; refer Table 1. The exergy loss or irreversibility is highest in this auxiliary boiler among the other auxiliary used on board ship. The major observations relating to the reduction of irreversibility in auxiliary boiler is as follows [7]:

(a) The chemical exergy identified is $'E^0 = 45{,}704$ kJ/kg for the light diesel oil used for combustion purpose.

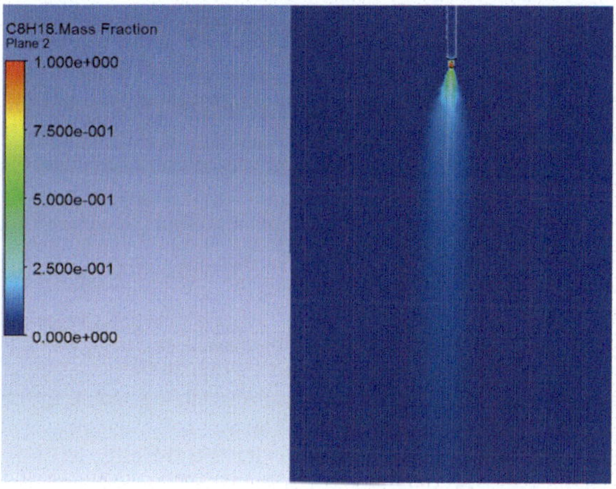

Fig. 4 C_8H_{18} mass fraction of the auxiliary boiler

Table 1 Irreversibility and its remedies in the auxiliary boiler [8]

S. No.	Components	Irreversibility generation	Remedial action
1	Cascade tank	Temperature difference of cold feed water and hot condensate water	Pre heating of feed water
2	Economizer and optimizer	Temp difference, non uniform gradient Pressure drop and low heat transfer	Use as low as possible driving force Use as uniform gradients Reduce number of baffles Optimize flow velocity
3	Steam condenser	Thermal difference Air in system.	Use of nucleate boiling Purging to be followed
4	System valves	Pressure drop	Fully opened or shut
5	Furnace	A chemical reaction for oxidation of the light diesel oil An internal heat transfer between high temperature products and unburned reactants A physical mixing process	Preheating of air Use as low driving force as possible Mix it as uniformly as possible Make it as gradually as possible
6	Blower	High air temperature	Use of intercooler

(b) The increase in unavailable energy is 11.689×10^4 J/kg by preheating of the combustion air.

(c) The auxiliary boiler $N_{2\text{nd law}}$ efficiency is = 0.3740 and irreversibility is I/M_{H_2O} = 870.5189 kJ/kg H_2O.

(d) Chemical reaction and physical transport processes are the sources of irreversibility in combustion. All chemical reaction is thermodynamically reversible.

References

1. Moran MJ (1982) Availability analysis: a guide to efficient energy use. Prentice Hall, Inc., Englewood Cliffs, New Jersey
2. Moran MJ, Sciubba E (1994) Exergy analysis: principles and practice. J Eng Gas Turbines Power 116:285
3. Kotas TJ (2012) The exergy method of thermal plant analysis. Exergon Publishing Company UK Ltd., London. ISBN: 978-1-908341-89-1
4. Zhao Z, Yuan Z, Cui Q, Yan N (2010) Exergy analysis of a WGZ220/6.8-1 type alkali recovery boiler. Mod Appl Sci 4(2):23–29
5. Pal MK, Kumar A, Chandra H (2013) Energy and exergy of boiler and turbine of coal fired thermal power plant. Int J Eng Res Technol 2(6). ISSN: 2278-0181
6. Som SK, Datta A, Enis T (2007) Thermodynamic irreversibility and exergy balance in combustion process. Prog Energy Combust Sci 34:351–376
7. Akubue GU, Enibe SO, Nijoku HO, Unachukwu GO (2014) Exergy analysis of a steam boiler plant in a brewery in Nieria. Int J Sci Eng Res 5(5):868–887
8. Ohijeagbon IO, Waheed MA, Jekayinfa SO, Lasode OA (2014) Exergetic modelling of oil-fired steam boilers. Niger J Technol (NIJOTECH) 33(4):523–536

Jitendra Singh Pal is Assistant Professor at Tolani Maritime Institute, Pune. After serving for 20 years in Indian Navy, he is currently pursuing PhD from Visvesvaraya Technological University, Belagavi.

S. N. Sapali is Professor of mechanical engineering at College of Engineering, Pune, and is pursuing PhD from Indian Institute of Technology, Kharagpur. He is the author of Refrigeration and Air Conditioning published by Prentice Hall of India. He is the Chairman and a member of expert committees for opening new MBA, MCA, pharmacy and engineering institutes under AICTE, New Delhi.

T. R. Anil is Professor of aeronautical engineering at Gogte Institue of Technology, Belagavi. His main areas of interest were applied thermodynamics and internal combustion engine, obtaining a PhD for his work in the former subject.

Thermoeconomic Modelling and Analysis of Energy Conversion System: Intercooled Recuperated Gas Turbine

Mithilesh Kumar Sahu, Tushar Choudhary and Sanjay

Abstract Global increase in population growth and improved living standards leads to an increase in energy demand that too at a lowest possible cost. This increased rate of energy demand challenges the energy engineers for effective energy production at minimum possible cost. In this regard, an analysis of power utilities from both perspectives, i.e. fuel efficiency and cost effectiveness becomes necessary to predict the thermodynamic and economic performance of energy systems. In the present work, intercooled recuperated gas turbine cycle-based power utility has been conceptualized and the same has been thermoeconomically analysed by using the 'Average Cost Theory' approach. The present article deals with the mathematical modelling of each and every component of cycle and the so obtained results are analysed on the basis of effect of key operating parameters (compressor pressure ratio, turbine inlet temperature and ambient temperature) on thermodynamics and economics of the investigating power utility. The analysis reports that during summer seasons, the decline in plant-specific work can be seen up to 10% for increase in ambient temperature by 20 °C, while for base case parameters ($r_{pc} = 30$, TIT = 1500 K, $\eta_{AC} = 88\%$ and $\eta_{GT} = 90\%$), the plant-specific work, total cost rate and exergetic efficiency for investigating cycle being are 600 kJ/kg, 23.61 \$/h and 48.29%, respectively.

Keywords Air film cooling · Exergy analysis · Gas turbine · Total cost rate Thermoeconomic analysis

M. K. Sahu (✉)
Mechanical Engineering Department, GVP College of Engineering (Autonomous), Visakhapatnam, India
e-mail: mithleshsahu3646@gmail.com

T. Choudhary
School of Mechanical and Building Science, VIT Bhopal University, Kotrikalan, Sehore 466114, India

Sanjay
Mechanical Engineering Department, National Institute of Technology Jamshedpur, Jamshedpur, India

J. Chattopadhyay et al. (eds.), *Renewable Energy and its Innovative Technologies*, https://doi.org/10.1007/978-981-13-2116-0_7

Nomenclature

c	Cost per unit of exergy ($/MJ)
$\dot{C}$	Exergetic cost flow rate ($/h)
c_p	Specific heat at constant pressure (kJ/kgK)
$\dot{C}_D$	Cost rate of exergy destruction ($/h)
$\dot{C}_L$	Cost rate of exergy loss ($/h)
$\dot{C}_T$	Total cost flow rate ($/h)
$\dot{C}_F$	Cost rate of fuel exergy of component ($/h)
C_F	Average fuel cost per exergy unit of component ($/MW)
$\dot{C}_P$	Cost rate of product exergy of component ($/h)
C_P	Average product cost per exergy unit of component ($/MW)
$\dot{C}_f$	Cost rate of fuel supply ($/h)
c_f	Cost of fuel supply ($/MJ)
$\dot{E}$	Exergy flow rate (MW)
$\dot{E}_D$	Exergy destruction rate (MW)
F_{sa}	Correction factor to account actual blade surface
f	Exergoeconomic factor
G	Constant in cost equations
h	Specific enthalpy of the stream (kJ/kg)
LHV	Lower heating value of fuel (MJ/kg)
Ic	Intercooler
$\dot{m}$	Mass flow rate (kg/s)
N	Number of hours of plant operation per year
p	Pressure (bar)
p_0	Reference or ambient pressure (bar)
r_{pc}	Compressor pressure ratio
Rc	Recuperator
T	Temperature (K)
T_0	Reference or ambient temperature (K)
y_D	Exergy destruction ratio (%)
y_L	Exergy loss ratio (%)
Z	Purchase cost associated with the component ($)
$\dot{Z}$	Investment cost flow rate ($/h)

Greek Symbols

η_{AC}	Compressor isentropic efficiency (%)
η_{GT}	Gas turbine isentropic efficiency (%)
φ	Maintenance factor
f	Exergoeconomic factor
r	Relative cost difference

ε	Exergetic efficiency (%)
η	Efficiency (%)

Subscripts

0	Reference environment
a	Air
C	Compressor
CC	Combustion chamber
D	Destruction
e	Exit
f	Fuel
g	Gas
GT	Gas turbine
i	Inlet, stream
k	Component number
L	Loss

Acronyms

BGT	Basic gas turbine
C	Compressor
CC	Combustion chamber
CRF	Capital recovery factor
GT	Gas turbine
HPC	Higher pressure compressor
IcRcGT	Intercooled recuperated gas turbine
LPC	Lower pressure compressor
TIT	Turbine inlet temperature

1 Introduction

The starting work in thermoeconomic field has been cited during the 60s but noteworthy development and advancement took place at the early 90s. Thermoeconomic examination of energy systems turns out to be the center of attraction for researchers after the introduction of CGAM problem [1]. The CGAM problem is well-known gas turbine-based cogeneration problem in the field of thermoeconomic analysis, and the CGAM problem is named after the names of scientists Frangopoulos [2], Tsatsaronis [3], Valero [4] and Von Spakovsky [5] who compared their approach of thermoeconomic analysis and optimization for analysis of gas turbine-based cogeneration system. da Gama Cerqueira et al. [6] have thermoeconomically analysed the cogeneration cycle by four different practices and reported the importance of division of exergy in its components (chemical, mechanical and thermal). Kwak et al. [7] have reported the exergy and thermoeconomic analysis of power plants by the help of computer programme developed through modelling. Colpan and Yesin [8] have analysed the combined cogeneration systems with the help of energy, exergy and thermoeconomic analysis and conveyed the advantage of system with extraction condensing steam turbine compared to back-pressure steam turbine systems. Sahoo [9] reported the exergoeconomic analysis and also introduces an evolutionary programming technique of thermoeconomic optimization for cogeneration power cycle. In the field of thermoeconomic analysis, Kim [10] introduces the 'wonergy' concept of thermoeconomic analysis and the methodology has been validated with the help of CGAM and other cogeneration problems. Seyyedi et al. [11] have proposed a new technique for thermoeconomic optimization named as structural optimization technique and stated that it has a rapid convergence compared to other methodologies. Ahmadi et al. [12] reported an extensive exergy-based economic and environmental impact analysis, and also presents the optimization of several combined cycle power plants. Bakhshmand et al. [13] have exergoeconomically analysed and optimized a triple-pressure combined cycle power plant. Vandani et al. [14] reported the effects of fuel (natural gas and diesel) on environment in the operation of combined cycle power plants. On analysing the literature till date, it suggested that most of the investigators have adopted the same well-known CGAM problem for applying their methodologies. In view of this literature gap of thermoeconomic modelling and analysis of cooled complex, gas turbine-based power plants is obtained. In this regard, the presented work aims to thermoeconomically analyse the complex gas turbine cycle (Intercooled Recuperated Gas Turbine Cycle). The distinctive feature of this work is that complex gas turbine with blade cooling has been adopted for analysis which provides new directions to this field. In earlier work of author's [15–22], they have thermoeconomically investigated the effect of operating parameters on other complex gas turbine cycle-based utilities for different energy application. The schematic illustration of intercooled recuperated gas turbine featuring air film blade cooling is shown in Fig. 1.

2 Mathematical Modelling

Thermoeconomic investigation of IcRcGT cycle has been performed by modelling the various components of the cycle with the help of mass, heat, energy, exergy and cost balance equations. The 'Average Cost Theory' approach has been considered for cost calculation of each stream of the cycle from work of Seyyedi [11]. The operating parameters for the base case cycle have been taken from the author's earlier work [18] and presented in Table 1.

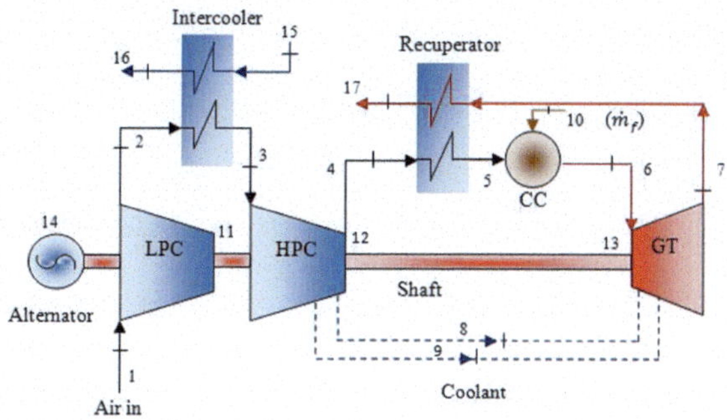

Fig. 1 Schematic illustration of intercooled recuperated gas turbine-based power utility

Table 1 Operating parameters for analysis of the proposed cycle

Parameter	Symbol	Adopted value	Unit
Gas properties	$c_p = f(T)$		kJ/kg K
	$h = \int_{T_0}^{T} c_p(T)dT$		kJ/kg
Compressor	Isentropic efficiency (η_{AC})	88	%
	Mechanical efficiency (η_{mech})	98.5	%
	Compressor pressure ratio	30	–
Intercooler	Pressure drop (air stream)	2.0% of p_{entry}	bar
	Pressure drop (water stream)	2.0% of p_{entry}	bar
	Effectiveness	0.92	–
Recuperator	Pressure drop (air stream)	5.0% of p_{entry}	bar
	Pressure drop (gas stream)	3.0% of p_{entry}	bar
	Effectiveness	0.75	–
Combustion chamber	CC efficiency (η_{CC})	99.5	%
	Pressure loss (p_{loss})	2.0% of p_{entry}	bar
	Lower heating value (LHV)	42.0	MJ/kg
	Fuel pressure	1.5 * p_{cc}	bar
Gas turbine	Isentropic efficiency (η_{GT})	90	%
	Exit pressure	1.08	bar
	Blade temperature of turbine (T_b)	1150	K
	Turbine inlet temperature (TIT)	1500	K

2.1 *Air/Gas Model*

The specific heat of air and flue gases is taken as temperature-dependent parameter and related polynomials for the same are given as under:

Polynomial for c_{pa} and c_{pg} taken from work of Touloukian and Tadash [23]

$$c_{pa} = 1.023204 - 1.76021 \times 10^{-4}T + 4.0205 \times 10^{-7}T^2 - 4.87272 \times 10^{-11}T^3 \tag{1}$$

$$\begin{aligned} c_{pg} = \big[&15.276826 + 0.01005T - 3.19216 \times 10^{-6}T^2 \\ &+ 3.48619 \times 10^{-10}T^3 + x_0\big(0.104826 + 5.54150 \times 10^{-5}T \\ &-1.67585 \times 10^{-8}T^2 + 1.18266 \times 10^{-12}T^3\big)\big]/V \end{aligned} \tag{2}$$

The important thermodynamic properties are evaluated as under

$$h = \int_{T_0}^{T} c_p(T)\mathrm{d}T \tag{3}$$

$$\theta = \int_{T_0}^{T} c_p(T)\frac{\mathrm{d}T}{T} \tag{4}$$

$$s = \theta - R\ \ln\left(\frac{p}{p_0}\right) \tag{5}$$

$$E = h - T_0 \cdot s \tag{6}$$

2.2 Compressor Model

The required equations to determine the unknown state points and to compute the performance are as under

$$p_{\text{optimim}} = \sqrt{p_i \cdot p_e} \tag{7}$$

$$T_e = T_i\left\{1 + \frac{1}{\eta_{AC}}\left[\left(\frac{p_e}{p_i}\right)^{\frac{\gamma_{a-1}}{\gamma_a}} - 1\right]\right\} \tag{8}$$

$$\dot{W}_{LPC} = \dot{m}_i \cdot (h_e - h_i) \tag{9}$$

$$\dot{W}_{HPC} = \dot{m}_e \cdot h_e + \sum \dot{m}_{\text{cool},j} \cdot h_{\text{cool},j} - \dot{m}_i \cdot h_i \tag{10}$$

where $m_{\text{cool},j}h_j$ denotes the heat contents of the coolant (air stream taken out from the compressor unit for blade cooling to avoid failure of blade material).

2.3 Intercooler Model

Implementation of intercooling technique in between the two compressors saves some compression work which ultimately results in an equivalent increase in plant-specific work. In this work, surface type intercooler is utilized to reduce the temperature of compressed air by exchanging heat with the cooling medium mostly water.

The effectiveness of intercooler is given by

$$\varepsilon_{\mathrm{ic}} = \frac{\left(T_{\mathrm{ic,a}}\right)_{\mathrm{in}} - \left(T_{\mathrm{ic,a}}\right)_{\mathrm{out}}}{\left(T_{\mathrm{ic,a}}\right)_{\mathrm{in}} - \left(T_{\mathrm{ic,w}}\right)_{\mathrm{in}}} \tag{11}$$

The energy balance gives

$$\dot{m}_{\mathrm{ic,a}} \cdot c_{\mathrm{p,a}} \cdot \varepsilon_{\mathrm{ic}}\left\{\left(T_{\mathrm{ic,a}}\right)_{\mathrm{in}} - \left(T_{\mathrm{ic,a}}\right)_{\mathrm{out}}\right\} - \dot{m}_{\mathrm{ic,w}} \cdot c_{\mathrm{p,w}}\left\{\left(T_{\mathrm{ic,w}}\right)_{\mathrm{out}} - \left(T_{\mathrm{ic,w}}\right)_{\mathrm{in}}\right\} = 0 \tag{12}$$

2.4 Recuperator Model

The effectiveness of recuperator is given by

$$\varepsilon_{\mathrm{Rc}} = \frac{\left(T_{\mathrm{Rc,a}}\right)_{\mathrm{out}} - \left(T_{\mathrm{Rc,a}}\right)_{\mathrm{in}}}{\left(T_{\mathrm{Rc,g}}\right)_{\mathrm{in}} - \left(T_{\mathrm{Rc,a}}\right)_{\mathrm{in}}} \tag{13}$$

The energy balance of recuperator gives

$$\dot{m}_{\mathrm{Rc,g}} \cdot c_{\mathrm{p,g}} \cdot \varepsilon_{\mathrm{Rc}}\left\{\left(T_{\mathrm{Rc,g}}\right)_{\mathrm{in}} - \left(T_{\mathrm{Rc,g}}\right)_{\mathrm{out}}\right\} - \dot{m}_{\mathrm{Rc,a}} \cdot c_{\mathrm{p,a}}\left\{\left(T_{\mathrm{Rc,a}}\right)_{\mathrm{out}} - \left(T_{\mathrm{Rc,a}}\right)_{\mathrm{in}}\right\} = 0 \tag{14}$$

2.5 Combustion Chamber Model

In combustion chamber, there are various types of losses. In this work, the losses are taken care by considering a pressure drop loss and supplied exergy loss during its operation.

$$\dot{m}_{\mathrm{e}} = \dot{m}_{\mathrm{i}} + \dot{m}_{\mathrm{f}} \tag{15}$$

$$\dot{m}_{\mathrm{f}} \cdot \mathrm{LHV} \cdot \eta_{\mathrm{cc}} = \dot{m}_{\mathrm{e}} \cdot h_{\mathrm{e}} - \dot{m}_{\mathrm{i}} \cdot h_{\mathrm{i}} \tag{16}$$

$$\dot{m}_{\mathrm{f}} = \frac{[\dot{m}_{\mathrm{i}} \cdot h_{\mathrm{e}} - \dot{m}_{\mathrm{i}} \cdot h_{\mathrm{i}}]}{[\eta_{\mathrm{cc}} \cdot \mathrm{LHV} - h_{\mathrm{e}}]} \tag{17}$$

$$\dot{m}_{\mathrm{g}} = \dot{m}_{\mathrm{a}} + \dot{m}_{\mathrm{f}} \tag{18}$$

$$p_{\mathrm{e}} = p_{\mathrm{i}}(1 - \Delta p_{\mathrm{cc}}) \tag{19}$$

where $\Delta p_{\mathrm{cc}} = 0.02$.

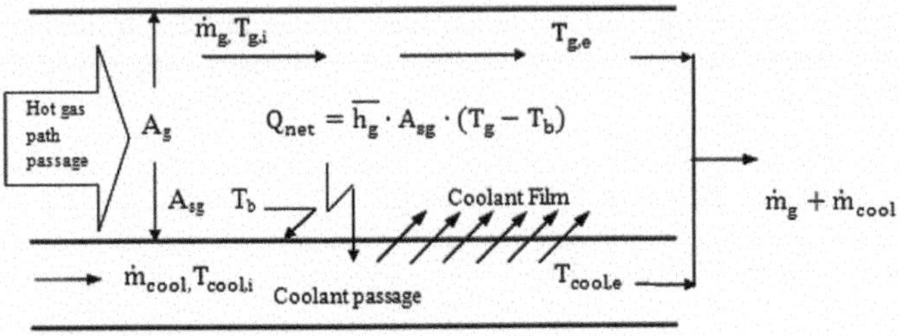

Fig. 2 Air film cooling of turbine blades

2.6 Cooled Gas Turbine Model

The mass of coolant bled from compressor ($\dot{m}_{cool}$) is given by Sanjay et al. [24] and schematic diagram for air film cooling has been presented in Fig. 2.

$$\zeta = \frac{\dot{m}_{cool}}{\dot{m}_g} = (1 - \eta_{iso,air})\frac{St_i \cdot S_g}{\varepsilon_{cool} \cdot t \cdot \cos\alpha} * \frac{c_{pg}(T_{g,i} - T_b)}{c_{p,cool}(T_b - T_{cool,i})} * F_{sa} \tag{20}$$

$$\dot{m}_e = \dot{m}_i + \sum \dot{m}_{cool,j} \tag{21}$$

$$T_e = T_i \left\{ 1 - \eta_{GT} \left[1 - \left(\frac{p_i}{p_e} \right)^{\frac{1-\gamma_g}{\gamma_g}} \right] \right\} \tag{22}$$

$$\dot{W}_{GT} = \left(\dot{m}_i h_i + \sum \dot{m}_{cool,j} h_{cool,j} - \dot{m}_e h_e \right) * \eta_{mech} \tag{23}$$

$$\dot{W}_{net} = (\dot{W}_{GT} - \dot{W}_{HPC} - \dot{W}_{LPC}) * \eta_{alternator} \tag{24}$$

2.7 Economic Model

In economic evaluation, the primary objective of the process is to identify the numerous sources of cost involved. In power plant capital investment in purchasing equipment, operating and maintenance cost and fuel used for operation are such cost sources [18].

Governing equations for PEC (Z) are shown in Tables 2 and 3.

$$\dot{Z}_k = \frac{Z_k \cdot \mathrm{CRF} \cdot \varphi}{N} \tag{25}$$

$$\dot{C}_f = c_f \cdot \dot{m}_f \cdot \mathrm{LHV} \tag{26}$$

$$\dot{C}_T = c_f \cdot \dot{m}_f \cdot \mathrm{LHV} + \sum_{k=1}^{n} \dot{Z}_k \tag{27}$$

Table 2 Equations for calculating the PEC (Z) for the components [11]

Component	Purchase cost equation
Compressor	$Z_{\mathrm{C}} = \left(\frac{G_{11}\cdot \dot{m}_{\mathrm{a,i}}}{G_{12}-\eta_{\mathrm{AC}}}\right)\left(\frac{p_{\mathrm{e}}}{p_{\mathrm{i}}}\right)\ln\left(\frac{p_{\mathrm{e}}}{p_{\mathrm{i}}}\right)$
Intercooler	$Z_{\mathrm{Ic}} = G_{51}\left(\frac{\dot{m}_{\mathrm{a,i}}\{(h_{\mathrm{Ic,a}})_{\mathrm{i}}-(h_{\mathrm{Ic,a}})_{\mathrm{e}}\}}{U(\Delta\mathrm{TLM})}\right)^{0.6}$
Recuperator	$Z_{\mathrm{Rc}} = G_{51}\left(\frac{\dot{m}_{\mathrm{g,i}}\{(h_{\mathrm{Rc,g}})_{\mathrm{i}}-(h_{\mathrm{Rc,g}})_{\mathrm{e}}\}}{U(\Delta\mathrm{TLM})}\right)^{0.6}$
Combustion chamber	$Z_{\mathrm{CC}} = \left(\frac{G_{21}\cdot \dot{m}_{\mathrm{a,i}}}{G_{22}-\frac{p_{\mathrm{e}}}{p_{\mathrm{i}}}}\right)(1+\exp(G_{23}\cdot T_{\mathrm{e}}-G_{24}))$
Gas turbine	$Z_{\mathrm{GT}} = \left(\frac{G_{31}\cdot \dot{m}_{\mathrm{g,e}}}{G_{32}-\eta_{\mathrm{GT}}}\right)\ln\left(\frac{p_{\mathrm{i}}}{p_{\mathrm{e}}}\right)(1+\exp(G_{23}\cdot T_{\mathrm{i}}-G_{24}))$

Table 3 Component-wise constants (Table 2) to determine purchase cost of components [11]

Component	Constants used in the purchase cost equation
Compressor	G_{11} =39.50 \$/kg/s; G_{12} =0.9
Intercooler	G_{51} =2290 \$/m$^{1.2}$; U =0.018 kW/(m$^{1.2}$ K)
Recuperator	G_{51} =2290 \$/m$^{1.2}$; U =0.018 kW/(m$^{1.2}$ K)
Combustion chamber	G_{21} =25.65 \$/kg/s; G_{22} =0.995; G_{23} =0.018 K^{-1}; G_{24} =26.4
Gas turbine	G_{31} =266.3 \$/kg/s; G_{32} =0.920; G_{33} =0.036 K^{-1}; G_{34} =54.4

2.8 *Cost Model*

The cost rate equations [25] for different streams are as below

$$\dot{C}_{\mathrm{i}} = c_{\mathrm{i}}\cdot \dot{E}_{\mathrm{i}} = c_{\mathrm{i}}(\dot{m}_{\mathrm{i}}\cdot e_{\mathrm{i}}) \tag{28}$$

$$\dot{C}_{\mathrm{e}} = c_{\mathrm{e}}\cdot \dot{E}_{\mathrm{e}} = c_{\mathrm{e}}(\dot{m}_{\mathrm{e}}\cdot e_{\mathrm{e}}) \tag{29}$$

$$\dot{C}_{\mathrm{w}} = c_{\mathrm{w}}\cdot \dot{W} \tag{30}$$

$$\dot{C}_{\mathrm{q}} = c_{\mathrm{q}}\cdot \dot{E}_{\mathrm{q}} \tag{31}$$

The generalized equation of cost assignment is as under

$$\sum_{\mathrm{e}} \dot{C}_{\mathrm{e},k} + \dot{C}_{\mathrm{w},k} = \dot{C}_{\mathrm{q},k} + \sum_{\mathrm{i}} \dot{C}_{\mathrm{i},k} + \dot{Z}_k \tag{32}$$

The auxiliary equations for both (compressor and expander) are taken from Tsatsaronis work [26].

2.9 *Performance and Evaluation Model*

The following governing equations are used to compute the thermoeconomics of presented cycle [25]:

Total cost rate

$$\dot{C}_{\mathrm{T}} = c_{\mathrm{f}} \cdot \dot{m}_{\mathrm{f}} \cdot \mathrm{LHV} + \sum_{i=1}^{6} \dot{Z}_k \tag{33}$$

Exergetic efficiency

$$\varepsilon = \frac{\dot{E}_{\mathrm{P}}}{\dot{E}_{\mathrm{F}}} \times 100 \tag{34}$$

Average product cost per exergy unit of component

$$c_{\mathrm{P},k} = \frac{\dot{C}_{\mathrm{P},k}}{\dot{E}_{\mathrm{P},k}} \tag{35}$$

Average fuel cost per exergy unit of component

$$c_{\mathrm{F},k} = \frac{\dot{C}_{\mathrm{F},k}}{\dot{E}_{\mathrm{F},k}} \tag{36}$$

Cost rate of exergy destruction

$$\dot{C}_{\mathrm{D},k} = c_{\mathrm{F},k} \cdot \dot{E}_{\mathrm{D,Tot}} \tag{37}$$

Relative cost difference

$$r_k = \frac{c_{\mathrm{P},k} - c_{\mathrm{F},k}}{c_{\mathrm{F},k}} \tag{38}$$

Exergoeconomic factor

$$f_k = \frac{\dot{Z}_k}{\dot{Z}_k + \dot{C}_{\mathrm{D},k}} \tag{39}$$

3 Result and Discussion

A computer programme in MATLAB software has been developed to determine the values of chosen exergoeconomic parameters (mass flow rate of fuel and blade coolant, plant-specific work, energy efficiency, exergetic efficiency and total cost

Table 4 State properties and exergy rate of IcRcGT cycle for base case (r_{pc} = 30, TIT = 1500 K, η_{AC} = 88% and η_{GT} = 90%)

Stream No.	Mass flow rate (kg/s)	Temperature (K)	Pressure (bar)	Enthalpy (kJ/kg)	Exergy (MW)
1	1	288	1.0132	0	0
2	1	501.67	5.6227	220.0910	0.2007
3	1	305.09	5.5103	17.1893	0.1436
4	0.9637	530.41	30.396	251.0934	0.3458
5	0.9637	678.49	28.8762	418.0740	0.4327
6	0.9923	1500	28.2986	1608.767	1.3033
7	1.0285	727.85	1.08	510.8627	0.1895
8	0.0266	530.41	30.396	251.0934	0.0095
9	0.0096	428.78	15.7605	143.3018	0.0024
10	0.0285	288	42.4480	42,000	1.2439
11	–	–	–	–	0.2200
12	–	–	–	–	0.2328
13	–	–	–	–	1.0628
14	–	–	–	–	0.6007
15	0.25	288	1.0132	62.7	0.0004
16	0.25	466.26	1.0132	807.8402	0.0492
17	1.0285	602.66	1.0476	264.2974	0.0901

rate). The exergoeconomic analysis of IcRcGT cycle is presented through different illustrative graphs and tables. Tables 4, 5 and 6 represents the results obtained from MATLAB code and have data related to stream report, exergy analysis and cost analysis, respectively.

Figure 3 shows the variation of exergetic efficiency and total cost rate with compressor pressure ratio for fixed TIT (1600 K) and with turbine inlet temperature for fixed r_{pc} (30). Figure 3 shows that with increase in r_{pc} the exergetic efficiency curve slightly rises and then for higher r_{pc} it shows decreasing trend because for higher r_{pc} the increase in fuel consumption due to intercooling dominates over saving in compressor power. The total cost rate increases with r_{pc}, as higher r_{pc} results in higher investment cost rate as well as higher fuel cost rate.

For TIT variation, exergetic efficiency shows increasing trend from 1500 to 1600 K temperature range, this behaviour of curve is a result of dominating effect of increase in specific work over increase in fuel requirement. The rising curve of cost rate is a result of higher investment cost to acquire advanced blade cooling and blade material for gas turbine as well as due to higher fuel cost rate.

Figure 4 depicts the variation of energy efficiency and plant-specific work for varying r_{pc} at fixed TIT and for varying TIT at fixed r_{pc}. The curve of energy efficiency shows the same trend as for exergetic efficiency with increasing r_{pc} due to the same reason as explained in Fig. 3. The increase in plant-specific work is a result of saving in compressor work due to the intercooling of air between the LPC and HPC.

Table 5 Exergy analysis of IcRcGT cycle ($r_{pc} = 30$, TIT = 1500 K, $\eta_{AC} = 88\%$ and $\eta_{GT} = 90\%$)

Component	Fuel exergy (MW) [$\dot{E}_F$]	Product exergy (MW) [$\dot{E}_P$]	Exergy destruction (MW) [$\dot{E}_D = \dot{E}_F - \dot{E}_P$]	Exergy loss (MW) [$\dot{E}_L$]	Exergy destruction ratio (%) $\left[y_D = \frac{\dot{E}_D}{\dot{E}_{D,Tot}}\right]$	Exergy loss ratio (%) $\left[y_L = \frac{\dot{E}_L}{\dot{E}_{L,Tot}}\right]$	Exergetic efficiency (%) $\left[\varepsilon = \frac{\dot{E}_P}{\dot{E}_F}\right]$
LPC	0.22	0.200	0.0193	0	4.12	0	91.19
Ic	0.05	0.048	0.0082	0	1.75	0	85.55
HPC	0.23	0.214	0.0186	0	3.97	0	91.98
Rc	0.09	0.086	0.0124	0	2.64	0	87.47
CC	1.67	1.303	0.3483	0.02	74.10	11.6	77.73
GT	1.12	1.062	0.0629	0.18	13.39	88.4	94.40
System	1.24	0.600	0.4700	0.21	100	100	48.29

Table 6 Cost analysis of IcRcGT cycle ($r_{pc} = 30$, TIT = 1500 K, $\eta_{AC} = 88\%$ and $\eta_{GT} = 90\%$)

Component	Fuel cost (\$/h) $\dot{C}_F$	Product cost (\$/h) $\dot{C}_P$	$\dot{C}_D$ (\$/h) $[c_F \cdot \dot{E}_D]$	$\dot{C}_L$ (\$/h) $[c_F \cdot \dot{E}_L]$	$\dot{Z}$ (\$/h)	$\dot{Z} + \dot{C}_D + \dot{C}_L$ (\$/h)	f (%) $\left[\frac{\dot{Z}}{\dot{Z}+\dot{C}_D+\dot{C}_L}\right]$	r (%) $\left[\frac{c_{P,k}-c_{F,k}}{c_{F,k}}\right]$
LPC	6.23	6.691	0.549	0	0.45	1.002	45.17	17.6
Ic	4.05	1.903	0.585	0	2.15	2.738	78.60	−45.4
HPC	6.60	7.241	0.529	0	0.44	0.977	45.87	19.2
Rc	2.49	3.959	0.312	0	1.45	1.772	82.34	81.0
CC	33.0	33.04	6.860	0.49	0.02	7.376	0.35	28.7
GT	28.5	30.12	1.595	0	1.80	3.396	53.03	11.9
System	17.2	21.79	10.43	0.49	6.34	17.26	36.73	206

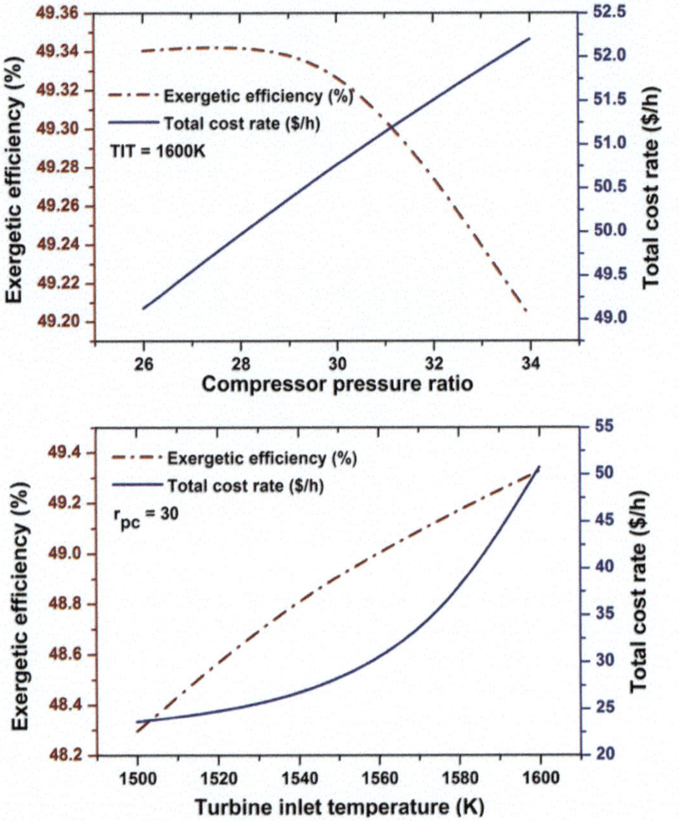

Fig. 3 Effect of r_{pc} and TIT on exergetic efficiency and cost rate

The specific work and energy efficiency variation with TIT for fixed r_{pc} is also presented in Fig. 4. The increase in TIT leads to higher plant-specific work. The curve of energy efficiency for varying TIT (fixed r_{pc}) resembles the trend of exergetic efficiency curve for varying TIT (fixed r_{pc}).

In the present work, efficiencies (energy and exergy efficiency) obtained are higher compared to conventional gas turbine cycles due to innovative cycle incorporating intercooling (which improves the plant-specific work) and recuperation (which tends to lower the fuel requirement). The chosen operating parameters are in line with what would be the best operating parameters for the IcRcGT cycle keeping in mind that the higher value of r_{pc} cannot be adopted.

Figure 5 depicts the variation of fuel and coolant flow rate with varying r_{pc} and TIT. On observing graph, it can be seen that as r_{pc} increases the fuel requirement get increased which is against the general concept of GT cycles. This nature of curve is due to the application of intercooler and recuperator, as in recuperator the exit stream of compressor is heated from turbine exhaust which temperature gets

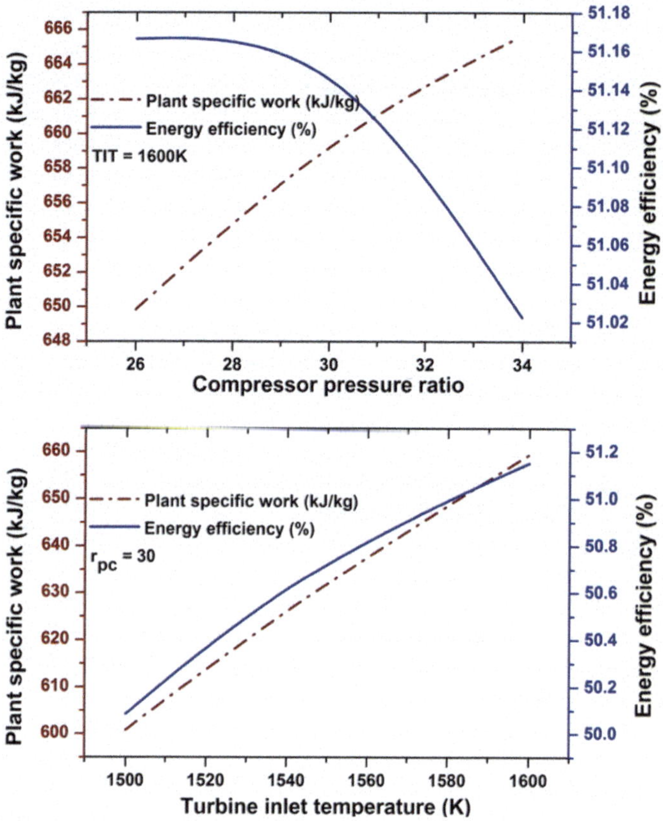

Fig. 4 Effect of r_{pc} and TIT on specific work and energy efficiency

decreased with increase in compressor pressure ratio. The higher r_{pc} results increase in coolant temperature which ultimately decreases the cooling effectiveness, so a higher amount of coolant is required for higher compressor pressure ratios which can be seen in shown Fig. 5. The variation of fuel and coolant flow rate for increasing TIT is in line with the concept of heat transfer as to get higher temperature more amount of heat need to be supplied which leads to higher fuel flow rate while higher TIT demands for more number of stages to be cooled with higher coolant flow rate.

Figure 6 depicts the effect of ambient temperature on exergetic efficiency and total cost rate. As ambient temperature increases the plant-specific work gets decreased while the fuel requirement also get decreased but due to the dominating effect of plant-specific work over lower fuel flow rate overall effect of decreasing efficiency has been achieved. In case of total cost rate, the saving in fuel results lower fuel cost rate while investment cost rate is increased for higher ambient temperatures. The overall effect formed for total cost rate is of decreasing nature.

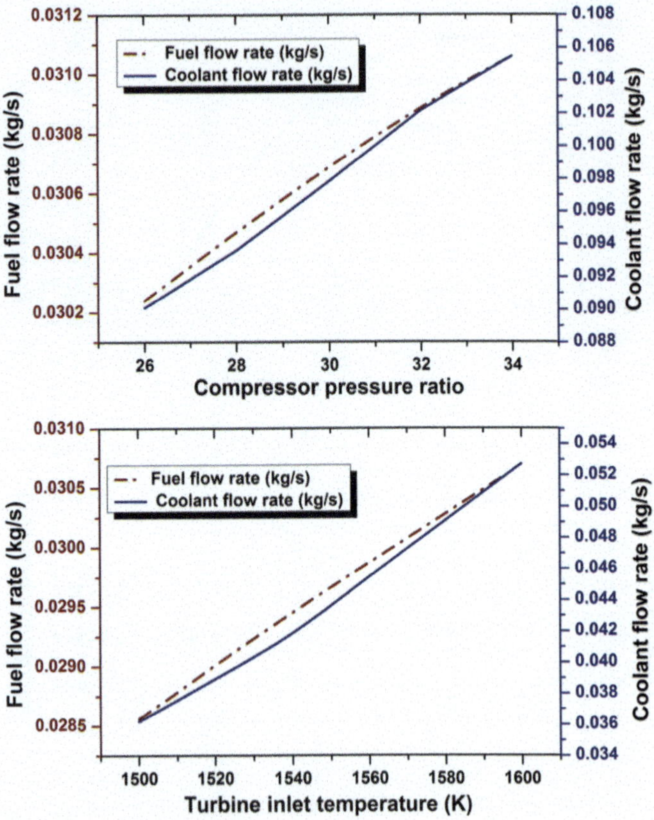

Fig. 5 Effect of r_{pc} and TIT on fuel and coolant flow rate

The higher ambient or reference (dead state) temperature results decrease in potential of doing work that is shown in Fig. 7 as ambient temperature increases the plant-specific work decreases. The curve of energy efficiency shows the similar trend as of exergy efficiency due to the dominating effect of decrease in plant-specific work over decease in fuel flow rate.

Figure 8 represents the variation of fuel and coolant mass flow rate with ambient temperature. On observing the figure, it can be seen that higher ambient temperature leads to lower fuel requirement as at the end of recuperation, the temperature is increased with increase in ambient temperature. The higher coolant requirement is a result of higher temperature of available coolant that leads to more amount of coolant to be bled for blade cooling due to lower cooling effectiveness.

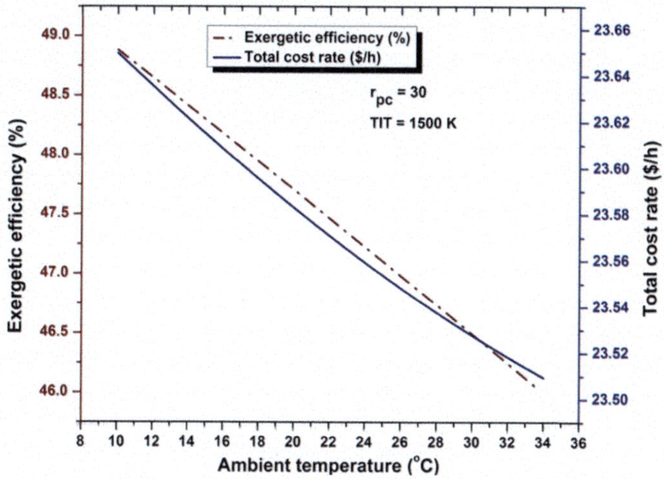

Fig. 6 Effect of ambient temperature on exergetic efficiency and total cost rate

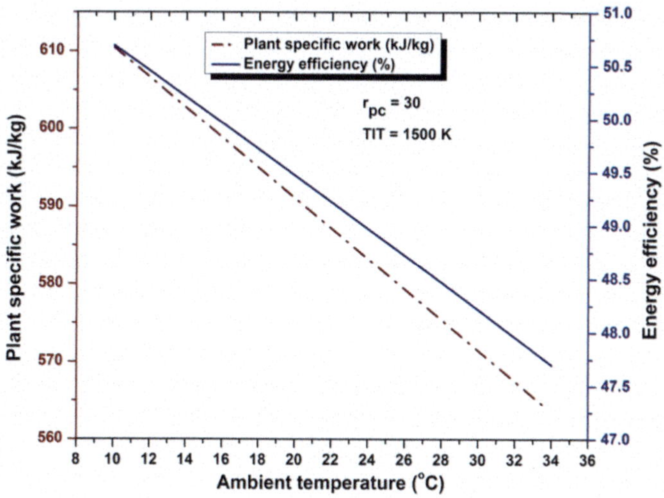

Fig. 7 Effect of ambient temperature on plant-specific work and energy efficiency

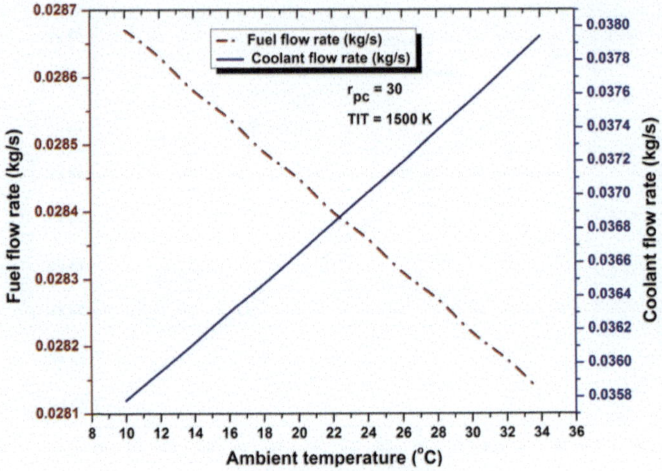

Fig. 8 Effect of ambient temperature on fuel and coolant flow rate

4 Conclusions

Based on thermoeconomic standard and average cost theory approach, IcRcGT-based power utility featuring air film cooling of turbine blade has been analysed and the important conclusions drawn from the analysis are as follows:

- The IcRcGT cycle for the base case (r_{pc} = 30, TIT = 1500 K, η_{AC} = 88% and η_{GT} = 90%) offers energy efficiency and exergetic efficiency as 50.08 and 48.29%, respectively.
- Total cost rate of intercooled recuperated GT cycle is 23.61 $/h for base case while the plant-specific work is 600 kJ/kg.
- The behaviour of thermoeconomic parameter (exergetic efficiency, total cost rate, plant-specific work, energy efficiency, fuel and coolant flow rate, etc.) with cycle operating parameters (compressor pressure ratio, turbine inlet temperature, ambient temperature, etc.) will turn out to be a useful tool for predicting the behaviour of performance parameter for given set of operating parameter.
- The deliberated article may become the helpful source for researchers working on GT-based power utilities as well as for power utility developers. The methodology presented here helps them to identify the cost formation resources and exergy destruction areas for basic as well as for complex cooled gas turbine cycles.

References

1. Valero A, Lozano MA, Serra L, Tsatsaronis G, Pisa J, Frangopoulos C, Von Spakovsky MR (1994) CGAM problem: definition and conventional solution. Energy 19:279–286

2. Frangopoulos C (1994) Application of thermoeconomic optimization methods to the CGAM problem. Energy 19:323–342
3. Tsatsaronis G, Pisa J (1994) Exergoeconomic evaluation and optimization of energy systems: application to the CGAM problem. Energy 19:287–321
4. Valero A, Serra L, Lozano MA, Torres C (1994) Application of the exergetic cost theory to the CGAM problem. Energy 19:365–381
5. Von Spakovsky MR (1994) Application of engineering functional analysis and optimization of the CGAM problem. Energy 19:343–364
6. da Gama Cerqueira SAA, Nebra SA (1999) Cost attribution methodologies in cogeneration systems. Energy Convers Manage 40:1587–1597
7. Kwak HY, Kim DJ, Jeon JS (2003) Exergetic, and thermoeconomic analyses of power plants. Energy 28:343–360
8. Colpan CO, Yesin T (2006) Thermodynamic, and thermoeconomic comparison of combined cycle cogeneration system. Int J Exergy 3(3):272–290
9. Sahoo PK (2008) Exergoeconomic analysis and optimization of a cogeneration system using evolutionary programming. Appl Therm Eng 28:1580–1588
10. Kim DJ (2010) A new thermoeconomic methodology for energy systems. Energy 35:410–422
11. Seyyedi SM, Ajam H, Farahat S (2010) A new approach for optimization of thermal power plant based on the exergoeconomic analysis and structural optimization method: application to the CGAM problem. Energy Convers Manage 51:2202–2211
12. Ahmadi P, Dincer I, Rosen MA (2011) Exergy, exergoeconomic and environmental analyses and evolutionary algorithm based multi-objective optimization of combined cycle power plants. Energy 36:5886–5898
13. Bakhshmand SK, Saray RK, Bahlouli K, Eftekhari H, Ebrahimi A (2015) Exergoeconomic analysis, and optimization of a triple-pressure combined cycle plant using evolutionary algorithm. Energy 93:555–567
14. Vandani AMK, Joda F, Boozarjomehry RB (2016) Exergic, economic and environmental impacts of natural gas and diesel in operation of combined cycle power plants. Energy Convers Manage 109:103–112
15. Sahu MK, Sanjay (2016) Investigation of the effect of air film blade cooling on thermoeconomics of gas turbine based power plant cycle. Energy 115:1320–1330
16. Sahu MK, Sanjay (2017a) Comparative exergoeconomic analysis of basic and reheat gas turbine with air film blade cooling. Energy 132:160–170
17. Sahu MK, Sanjay (2017b) Exergoeconomic investigation of power utility based on air film blade cooled gas turbine cycle. Appl Therm Eng 122:738–746
18. Sahu MK, Sanjay (2017c) Thermoeconomic investigation of power utilities: intercooled recuperated gas turbine cycle featuring cooled turbine blades. Energy 138:490–499
19. Sahu MK, Sanjay (2017d) Comparative exergoeconomics of power utilities: air-cooled gas turbine cycle and combined cycle configurations. Energy 139:42–51
20. Sahu MK, Sanjay (2018) Thermoeconomic investigation of basic and intercooled gas turbine based power utilities incorporating air-film blade cooling. J Cleaner Prod 170:842–856
21. Sahu MK, Sanjay, Choudhary Y (2016a) Thermoeconomic investigation of different gas turbine cycle configuration's for marine application. SAE technical paper 2016-01-2228. https://doi.org/10.4271/2016-01-2228
22. Sahu MK, Sanjay, Choudhary T (2016b) Exergoeconomic analysis of air cooled turboprop engine: air craft application. SAE technical paper 2016-01-2228. https://doi.org/10.4271/2016-01-2228
23. Touloukian YS, Tadash M (1970) Thermo-physical properties of matter. The TPRC Data Series. Vol-6 IFI/PLENUNM, New York, Washington
24. Sanjay Y, Singh O, Prasad BN (2008) Influence of different means of turbine blade cooling on the thermodynamic performance of combined cycle. Appl Therm Eng 28:2315–2326
25. Bejan A, Tsatsaronis G, Moran M (1996) Thermal design and optimization. Wiley, New York
26. Tsatsaronis G, Cziesla F (2002) Thermoeconomics. In: Encyclopedia of science and technology, 3rd edn, vol 16, pp 659–680

Mithilesh Kumar Sahu is working as Assistant Professor in the Department of Mechanical Engineering; GVP College of Engineering (A) Visakhapatnam, India. He submitted his Ph.D. in November 2017 in the Department of Mechanical Engineering NIT, Jamshedpur. His research is on exergoeconomic analysis of air film cooled complex gas turbine cycles and he published six SCI and five SCOPUS paper in international journals. He received his M.Tech from NIT, Jamshedpur, India in 2014 in thermal and fluids engineering stream.

Dr. Tushar Choudhary is working as Assistant Professor in School of Mechanical and Building Sciences, VIT University, Bhopal. He has completed his Ph.D. from National Institute of Technology, Jamshedpur in 2017. He has 6 years of teaching and research experience and published various papers in SCI/Scopus Indexed international journals. His area of interests is CFD, FEA, Solid oxide fuel cell, Thermodynamic Analysis of hybrid Gas Turbine cycle, Vibration, Turbomachinery.

Sanjay I am Professor Sanjay working in National Institute of Technology Jamshedpur as Professor of Mechanical Engineering. I did my Bachelor of Engineering (Mechanical) in 1993. Thereafter, I joined M.N.R. Engineering College, Allahabad, (now M.N.N.I.T, Allahabad) to continue my study by joining Master of Engineering in Design of Process Machines. I have industrial experience of around 6 years as inspection engineer specializing in inspection of steam turbines, process plant equipment, cross-country petroleum pipelines. I further decided to pursue higher studies and joined the Ph.D. programme at Uttar Pradesh Technical University, which was awarded in 2005. I have been engaged in pursuing cutting-edge research and have published many papers most of which are published in prestigious ASME International, Elsevier journals, SAE and other prestigious international level journals. Total citation of my 50+ research articles is 350+ and the i10 and h-index are 9 which are the highest in my institute.

Fresh Water Algae: A Best Option for Renewable Energy Generation

Neetu Singh and Binod Kumar Choudhary

Abstract Sustainable use of natural resources is key principle in present environment. It requires maximum utilization with minimum environmental impact. Use of Freshwater algae is better option for this. Freshwater algae are photosynthetic lower plants having 50% of oil content in their weight. By simple reactions, algae produce carbohydrates, lipids, and proteins in heavy stock over shortest period of time. These products are further converted into bioethanol, biodiesel, and many more useful products. This study deals with biofuel production from the microalgae collected from River Baldi, Doon Valley, Uttarakhand. Some common species are *Nitzschia* sp., *Chlorella* sp., *Cladophora* sp., *Zygnema* sp., *Spirogyra* sp., *Microcystis aeruginosa*, *Oscillatoria limnetica*, *Navicula* sp. 19.5% of lipid was obtained from 0.758 g dry wt m^{-2} microalgal biomass harvested from River Baldi by using soxhlet extraction method by using chloroform:methanol (2:1) as solvent. Extracted algal biomass was hydrolyzed in the form offermentable sugar which was converted into 58% bioethanol. Algal biomass was utilized for biofuel production, in biotreatment to treat waste water and as biofertilizer to enhance crop production. Hence, Microalgae have 3B role (Biofuel; Biotreatment; Biofertilizer). The results shows that biofuel derived from freshwater microalgae act as better substitute to fossil fuels in future.

Keywords Freshwater algae · Biofuel · River Baldi · Renewable energy

---

N. Singh
Department of Environmental Sciences, HNB Garhwal University (A Central University), Garhwal, Srinagar, Uttarakhand, India
e-mail: neetu.envirohnbgu@gmail.com

B. K. Choudhary (✉)
Department of Applied Physics, Cambridge Institute of Technology, Tatisilwai, Ranchi, Jharkhand, India
e-mail: binodvlsi@gmail.com

J. Chattopadhyay et al. (eds.), *Renewable Energy and its Innovative Technologies*,
https://doi.org/10.1007/978-981-13-2116-0_8

1 Introduction

Sustainable development of fuel production is requirement of present world. Reduction in the use of fossil fuels will reduces the amount of pollutants produced [1]. This target can be achieved by using renewable fuels. Scientific world moving towards the production of renewable alternative energy sources because fossil fuel going to be exhaust in future. Algal biofuel receives a lot of attention among the existing renewable alternatives to fossil fuels because 25% of global energy requirements could be supplied by algal biomass and biomass is also source of various food additives, chemicals and pharmaceuticals [2].

Algae are photosynthetic organisms that convert sunlight, water and carbon dioxide to algal biomass. Freshwater algae act as best raw material for energy generation because it has triple role: (a) activity of bioremediation of wastewater; (b) generating biomass for biofuel production; (c) act as biofertilizer in agricultural field. [3–6]. Recently, biofuel derived from algal biomass have gained much more interest than plant-based oils due to their high growth rate, high lipid content (20–75% by weight) than *Jatropha curcas* and rapeseed [7, 8] and growth of algal biomass requires very less land [9]. However, no attempt has been made so far on the freshwater algae collected from River Baldi. Therefore, the present study conducted on the production of biofuel from freshwater algae. This study will focus on

1. Analyzing water quality, collection of algal biomass and identification of algal species
2. Lipid extraction from the harvested algal biomass
3. Conversion of extracted lipid into biofuel
4. Release of fermentable sugars from the remaining residue of lipid-extracted biomass
5. Estimation of total yield of bioethanol (Biofuel).

2 Materials and Methods

2.1 Water Quality Measurement, Collection, and Identification of Algal Biomass

Freshwater algae biomass was collected from mountain River Baldi is a major tributary of the Song River in Doon Valley, Uttarakhand, India (Fig. 1) during the month of November 2016–February 2017, when the density and diversity of algal biomass is high. Water quality and Algal biomass was collected according to [10–14].

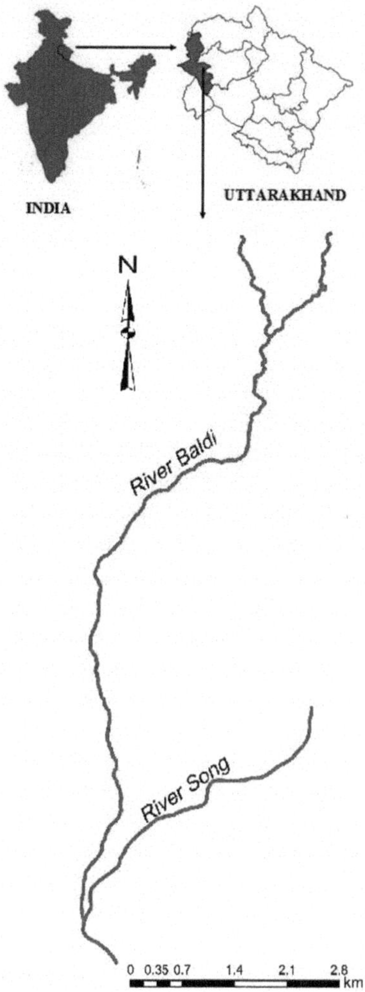

Fig. 1 Study area location map

2.2 *Lipid Extraction*

Bligh and Dyer [15] method was used for lipid analysis. Lipid extraction was done using soxhlet method. In this method 25 g of collected algal biomass was analyzed with 250 ml of chloroform: methanol (2:1) for 6 h in soxhlet system. A biphasic medium of water and chloroform was formed. The layer of chloroform containing extracted lipids was carefully collected upon volatilization of chloroform. The weight of oily extract was weighed and counted as oil content (% DW).

2.3 *Transesterification, Fatty Acid Profile, and Biofuel Properties*

Transesterification of extracted lipids into fatty acid methyl esters (FAMEs) done by methanolic sulphuric acid (6%) [16]. The FAMEs were analyzed using gas chromatography-mass spectroscopy (GC-MS) and using helium (1 mL/min) as carrier gas. 1 μL of lipid sample was injected and analyzed [9].

2.4 *Acid Hydrolysis and Sugar Fermentation*

Single-step hydrolysis method used for acid hydrolysis. 25 g lipid-extracted algal biomass was mixed with 250 mL of HCl and H_2SO_4 at 3 (1, 5, and 10%) different concentration. Hydrolysates were neutralized with 10M NaOH to pH 7 after cooling. Centrifugation of hydrolysates was done at 8000 rpm for 15 min. The supernatant was used for sugar and other analyses. For the fermentation, 0.5 g of dry yeast was added as inoculum to 100 mL of supernatant. Fermentation was carried out for 7 days at 32 °C and pH 5 with 150 rpm. Product further used for determination of Bioethanol concentration by colorimetrically using potassium dichromate method [17]. The maximum theoretical ethanol yield was calculated.

The maximum theoretical ethanol yield was calculated as follows:

$$Y_{\max}(\%) = \frac{\text{Ethanol produced in reactor(g)} \times 100}{\text{Initial sugar in the reactor(g)} \times 0.511}$$

3 Result and Discussion

3.1 *Water Quality, Algal Biomass Quantity, and Identification of Algae*

The water quality of any aquatic ecosystem indicates the density, diversity, and composition of biotic components of that ecosystem [18, 19]. Water quality of River Baldi is given in Table 1. Maximum dissolved oxygen was recorded during November to February (Winter season). It may due to high photosynthetic rate of algal communities that result in higher dissolved oxygen which results in high productivity of algal biomass [20]. *Chlorella vulgaris*, *Oscillatoria limnetica*, *Microcystis aeruginosa*, *Nitzschia* sp. *Spirogyra* sp., *Cladophora* sp., *Zygnema* sp. were the major algal species that have biofuel generation potential (Table 2) were identified during the study period. Total 0.758 g dry wt m^{-2} of algal biomass harvested from River Baldi to estimation of biofuel production potential.

Table 1 Water quality of Baldi River, Doon Valley (mean ± standard deviation obtained) during the period from November 2016–February 2017

Parameters	Water quality data
Water temperature (°C)	13.38 ± 2.09
Transparency (m)	0.17 ± 0.11
Water velocity (m s^{-1})	0.29 ± 0.13
Conductivity (mS cm^{-1})	0.40 ± 0.03
Turbidity (NTU)	36.52 ± 33.42
pH	7.65 ± 0.24
Dissolved oxygen (mg L^{-1})	8.51 ± 1.32
Alkalinity (mg L^{-1})	52.25 ± 9.09
Chlorides (mg L^{-1})	11.98 ± 1.75
Calcium (mg L^{-1})	164.00 ± 17.36
Magnesium (mg L^{-1})	81.58 ± 16.47
Total hardness (mg L^{-1})	245.58 ± 33.21
Nitrates (mg L^{-1})	0.08 ± 0.08
Phosphates (mg L^{-1})	0.04 ± 0.02
Sodium (mg L^{-1})	9.50 ± 0.88
Potassium (mg L^{-1})	3.91 ± 0.41

Table 2 Oil content of harvested algal species on dry weight basis

Algal species	Oil content (% dry weight)	Reference
Chlorella vulgaris	28–32	[21]
Oscillatoria limnetica	11	[22]
Microcystis aeruginosa	30	[22]
Nitzschia sp.	45–47	[21]
Cladophora sp.	3.5	[23]
Zygnema sp.	8.3	[23]
Spirogyra sp.	7.2	[23]

3.2 *Characterization of Fatty Acid Methyl Esters (FAMEs)*

Triglyceride (extracted lipids) was converted into FAMEs (Fig. 2). Total amount of 19.5% of lipid was obtained from the harvested freshwater algal biomass. The FAMEs were analyzed using gas chromatography-mass spectroscopy (GC-MS) (Fig. 3) revealed that major fatty acids that was obtained were saturated fatty acid (C16:0). Some of mono saturated fatty acids were also identified (C16:1, C18:1, and C20:1) During soxhlet extraction, thermo-degradation of long chain polyunsaturated fatty

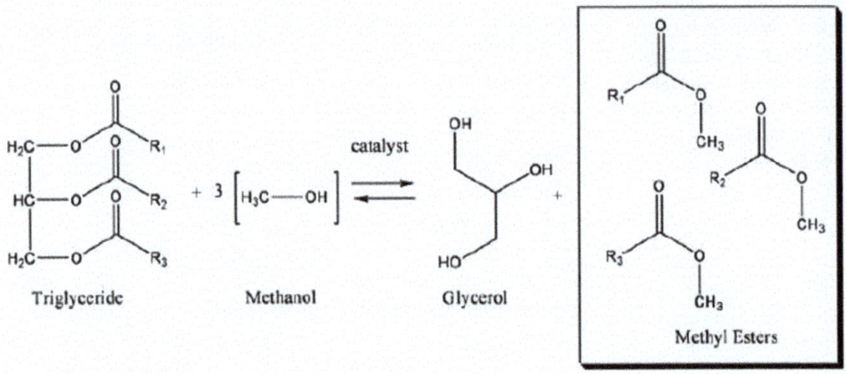

Fig. 2 Transesterification process (conversion of extracted lipids into FAMEs)

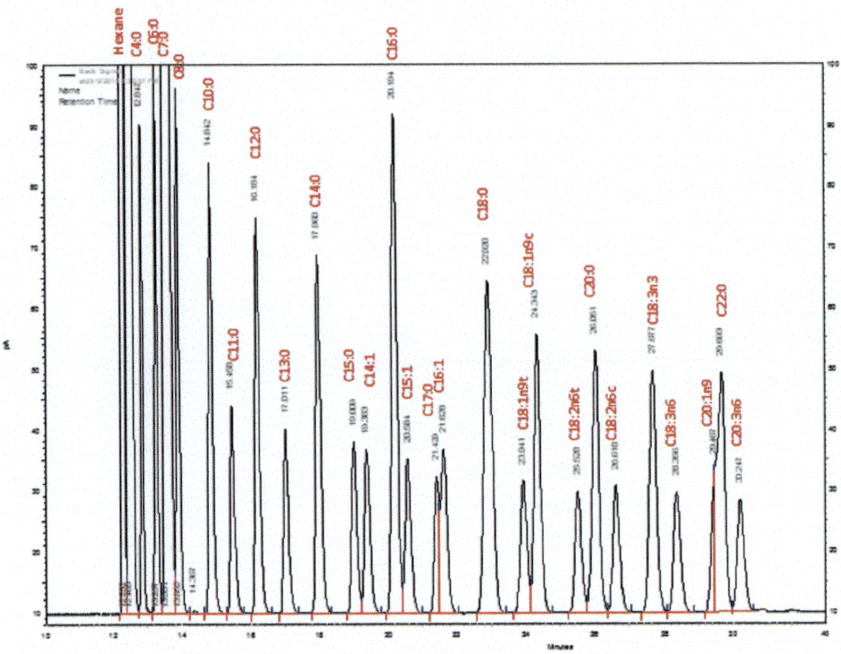

Fig. 3 GC-MS analysis of fatty acid methyl esters (FAMEs)

acids takes place resulting in number of fatty acids [24]. Fatty acids (C16:1; C18:1 and C18:2) are considered as major substitute for algal biofuel production [25]. Major composition of fatty acids were represented in Table 3.

Table 3 Fatty acid composition of wild algal biomass by using different solvents

Solvent	Fatty acid composition (wt%)			
	C16:0	C16:1	C18:1	C20:1
Chloroform:Methanol (2:1)	45	12	20	4

Table 4 Release of fermentable sugars by single-step hydrolysis

% of acid used	HCL (mean ± S.D.)	H_2SO_4 (mean ± S.D.)
1	11.02 ± 0.5	12.08 ± 0.8
5	68.14 ± 1.2	86.08 ± 2.1
10	206.18 ± 2.2	228.06 ± 0.5

*Weight of sugars = mg/g algae [26]

3.3 Release of Fermentable Sugars

Single-step hydrolysis method used for acid hydrolysis. Lipid-extracted algal biomass was further utilized for release of fermentable sugars when mixed with 250 ml of HCl and H_2SO_4 at 3 (1, 5 and 10%) different concentration (Table 4).

H_2SO_4 and HCl are considered as powerful agents for hydrolysis of cellulose [26]. The total yield of 19.5% of lipid was calculated from dry algal biomass. Conversion of algal biomass into fermentable sugars was estimated to 38.5%. Hence, total yield of bioethanol was recorded 58% of total harvested algal biomass.

4 Conclusion

This paper deals with the concept of generation of renewable energy used in the production of biofuel from freshwater algal biomass. 19.5% of lipid was extracted from algal biomass. Further, fermentable sugars converted into biofuel. The theoretical yield of conversion of total harvested algal biomass to bioethanol was estimated and found to be 58.0%. Algae serves as better tool for biofuel generation because they are economical, effective and easy growing. Algae fixed carbon dioxide from atmosphere that lowers CO_2 level, used for waste water treatment as they absorbs harmful pollutants which can be removed from water by harvesting them, do not need entire land and extra mechanical work for their growth. In agricultural country like India, algae considered as best substitute for replacing fossil fuel into renewable fuel.

References

1. Kirtay E (2009) The role of renewable energy sources in meeting Turkey's electrical energy demand. Energy Educ Sci Technol Part A 23:15–30
2. Briens C, Piskorz J, Berruti F (2008) Biomass valorization for fuel and chemicals production—a review. Int J Chem Reactor Eng 6:1–49
3. Mulbry W, Kondrad S, Pizarro C, Kebede-Westhead E (2008) Treatment of dairy manure effluent using freshwater algae: algal productivity and recovery of manure nutrients using pilot-scale algal turf scrubbers. Bioresour Technol 99:8137–8142
4. Olguın EJ (2003) Phycoremediation: key issues for cost-effective nutrient removal processes. Biotechnol Adv 22:81–91
5. Pizarro C, Mulbry W, Blersch D, Kangas P (2006) An economic assessment of algal turf scrubber technology for treatment of dairy manure effluent. Ecol Eng 26:321–327
6. Munoz R, Guieysse B (2008) Algal–bacterial processes for the treatment of hazardous contaminants: a review. Water Res 40:2799–2815
7. Amaro HM, Guedes AC, Malcata FX (2011) Advances and perspectives in using microalgae to produce biodiesel. Appl Energy 88(10):3402–3410
8. Schenk P (2008) Second generation biofuels: high efficiency microalgae for biodiesel production. Bioenergy Res 1:20–43
9. Arora N, Patel A, Pruthi PA, Pruthi V (2016) Synergistic dynamics of nitrogen and phosphorous influences lipid productivity in *Chlorella minutissima* for biodiesel production. Bioresour Technol 213:79–87
10. Ward JV (1974) A temperature stressed stream ecosystem below a Hypolimnial release Mountain Reservoir. Arch Hydrobiol 74:247–275
11. Wetzel RG, Likens GE (1991) Limnological analyses, 2nd edn. Springer, New York, pp 1–175
12. APHA (American Public Health Association) (2005) Standard methods for the examination of water and waste water. Washington, DC., pp 1–1368
13. Munshi JD, Roy SP, Munshi JS (2010) Manual of freshwater biota. Narendra Publishing House, Delhi, India, pp 1–435
14. Sharma RC, Singh N, Chauhan A (2016) The influence of physico-chemical parameters on phytoplankton distribution in a head water stream of Garhwal Himalayas: a case study. Egypt J Aquat Res 42(1):11–21
15. Bligh EG, Dyer WJ (1959) A rapid method of total lipid extraction and purification. Can J Biochem Physiol 37:913–917
16. Kumar V, Nanda M, Verma M (2017) Application of agar liquid-gel transition in cultivation and harvesting of microalgae for biodiesel production. Bioresour Technol 243:163–168
17. Crowell EA, Ough CS (1979) A modified procedure for alcohol determination by dichromate oxidation. Am J Enol Viticult 30:61–63
18. Hutchinson GE (1957) A treatise in limnology. Geography, Physics and Chemistry. John Wiley and Sons Inc., New York, USA 1957; I:1–1015
19. Ruttner F (1963) Fundamentals of limnology, 3rd edn. University Toronto Press, Toronto, pp 1–295
20. Sharma RK, Rathore V (2000) Pollution ecology with reference to commercially important fisheries prospects in rural-based water body: the Lake Sarsai Nawar, Etawah (Uttar Pradesh). Pollut Res 19:641–644
21. Chisti Y (2007) Biodiesel from microalgae. Biotechnol Adv 25:294–306
22. Abdo SM, Ahmed E, El-Enin SA, El Din RS et al (2014) Qualitative and quantitative determination of lipid content in microalgae for biofuel production. J Algal Biomass Utilization 5(3):23–28

23. Sawarkar LL, Nandkar PB (2013) Lipid content of some green algae for biodiesel. J Harmonized Res Appl Sci 1(3):77–79
24. Cheung PCK, Leung AYH, Ang PO (1998) Comparison of supercritical carbon dioxide and soxhlet extraction of lipids from a brown seaweed *Sargassum hemiphyllum* (Turn.), C Ag. J Agric Food Chem 46:4228–4232
25. Halim R, Gladman B, Danquah M, Webley P (2011) Oil extraction from microalgae for biodiesel production. Bioresour Technol 102:178–185
26. Sun Y, Cheng J (2002) Hydrolysis of lignocellulosic materials for ethanol production: a review. Bioresour Technol 83:1–11

Dr. Neetu Singh Research Scholar, Department of Environmental Sciences, HNB Garhwal University (A Central University) Srinagar-Garhwal, Uttarakhand. He is Ph.D. from same University Her research area Limnology specially fresh water.

Binod Kumar Choudhary Assistant Professor in Department of Applied Physics, Cambridge Institute of Technology, Ranchi. He is Ph.D. Pursuing from Vinoba Bhave University, Hazaribag, Jharkhand. His research area is Quantum optics.

Parametric Optimization of Surface Roughness and Overcut in Electric Discharge Machining of Al-SiC Using Copper Electrode

Sambeet Kumar Sahu, Subhasree Naik, Sudhansu Ranjan Das and Debabrata Dhupal

Abstract Nowadays, Al-SiC metal matrix composites are replacing conventional materials in aerospace and automobile applications because of their superior characteristics. However, present manufacturers are facing various challenges during machining these advanced, hard-to-cut materials. One alternative method suitable in such situation is electrical discharge machining which employs series of discrete sparks that melt and vaporize the material irrespective of the hardness. The objective of present experimental investigation is to estimate and develop comprehensive mathematical models for selected performance parameters such as, surface roughness (R_a), and overcut (OC) with the five input machining parameters such as low voltage current, high voltage current, pulse-on time, pulse-off time and flushing pressure through response surface methodology. Thereafter, the significance of the developed models have been verified through analysis of variance. Results showed surface finish improves with the increasing pulse-off time and flushing pressure whereas, overcut decreases with the increasing high voltage current but reverse effect obtained with pulse-on time. In end of the discussion, a multi-objective particle swarm optimization (MOPSO) algorithm has been employed for simultaneous optimization of muli-responses which provides helpful guidance for governing the machining parameters to enhance accuracy of the electrical discharge machined components and a confirmation test is performed at the optimized parameter setting for copper and brass electrodes respectively to validate the result.

Keywords Al-SiC metal matrix composite · Surface roughness · Overcut
ANOVA · PSO

S. K. Sahu · S. Naik · S. R. Das (✉) · D. Dhupal
Department of Production Engineering, Veer Surendra Sai University of Technology, Burla 768018, India
e-mail: das.sudhansu83@gmail.com

J. Chattopadhyay et al. (eds.), *Renewable Energy and its Innovative Technologies*,
https://doi.org/10.1007/978-981-13-2116-0_9

1 Introduction

Electro discharge machining (EDM), perhaps is the oldest and most widely used non-traditional machining process for electrically conductive materials which are very difficult-to-cut in traditional machining methods. In recent days, Al-SiC metal matrix composite is an imperative engineering material, used to attain the high performance because of its high stiffness and strength to weight ratio in addition to high temperature resistance. Hence, it has been attracted many researchers and engineers to investigate on it, due to its growing ranges of applications in the field of aerospace, military, die making industries and sports equipment. Machining of MMCs like Al-SiC composites on EDM is challenging due the presence of nonconductive and high melting point reinforcement particles that slows down the process and introduces additional complexity [1]. In recent past attempts are made to evaluate the machinability properties of Al-SiC by EDM process.

When the manufactures are dealing with multiple conflicting objectives, modelling technique helps in enhancing the efficacy of machining process. Although some theoretical models require simplifications, assumptions and approximations for approaching real machining process, do not consider any undesirable deficiency in the process. Therefore, analytical solutions cannot be easily extended to practical usage [2], and for this reason adequate modelling is essential to do quality predictions in a function of operating conditions. The model development by RSM and ANN are convenient methods for product as well as process improvement and have received considerable attention by the researchers in the last two decades.

Moreover, owing to the complicated behaviour of the machining processes, where a few distinctive and opposing goals must be optimized, at the same time the mono-objective optimization techniques don't allow to find the comprehensive ideal cutting conditions value which fulfils all the execution attributes in machining; hence the multi-objective optimization has turned into an undeniably vital and challenging assignment [3–6]. For sure, it offers most prominent measure of data with a specific end goal to settle on a choice on choosing process parameters in machining process. Amid all the optimization techniques, the particle swarm optimization (PSO) has explored as an effective and reliable tool in advanced computing technology for the outline of high-quality frameworks as it gives a straightforward, proficient, and well-organized way to optimize output, for example, performance, cost, and quality. The optimum parametric combinations for EDM process by approaching with PSO technique would help to accomplish a common goal of improving machining performance. A number of investigators have been employed methods which include statistical and analytical approach using RSM [7–14] for mathematical modelling in order to predict the responses and PSO [15–19] for multi-response optimization in order to control the process parameters during electric discharge machining process. The present research focused on process modelling and multi-response optimization during electrical discharge machining of Al-SiC MMC using statistical approach such as response surface methodology (RSM) followed by, computational approach like particle swarm optimization (PSO).

Table 1 Process parameters and levels

Control parameters	Symbols	Units	Levels		
			−1	0	+1
Low voltage current	A	A	5	10	15
High voltage current	B	A	1	1.5	2
Pulse-on time	C	μs	100	200	300
Pulse-off time	D	μs	10	20	30
Flushing pressure	E	kg/cm^2	0.2	0.4	0.6

2 Experimental Procedure

To accomplish the objective of the proposed research work low voltage current, high voltage current, pulse-on time, pulse-off time and flushing pressure are considered as major process variables, whereas the technological response characteristics are surface roughness and overcut. The different process parameters and their levels are shown in Table 1. Employing the selected parameters (five) with of each having three different levels, a well-sequential design layout was established based on Box-Behnken design of experiments (BBDOEs) in order to perform the EDM operation. For sequence of experiments, a design layout with corresponding experimental results are presented in Table 2.

The computer numerical control electrical discharge machine (EDM), model: ECOWIN PS 50ZNC with positive polarity for electrode and servo head (constant gap) was utilized to perform the experimental trials. Commercial grade EDM oil (paraffin oil) was used as dielectric fluid. The machining is executed on Al-SiC metal matrix using a cylindrical shape of 9 mm diameter copper as electrode material. The elemental composition of the workpiece material (here, Al-SiC) is presented in Table 3. The surface finish (Ra) followed by, radial overcut (OC) of the machined component are measured by utilizing roughness tester (make: Mitotoyo, model: Surftest SJ-210) and coordinate measuring machine (make: ZEISS, model: MC850), respectively. The schematic view of experimental work and methodology proposed in the current study, is presented in Fig. 1.

Table 2 Experimental design matrix and measured response values using copper electrode

Run order	Coded value					Actual setting					Responses	
	A	B	C	D	E	A (A)	B (A)	C (μs)	D (μs)	E (kg/cm^2)	R_a (μm)	OC (mm)
1	−1	−1	0	0	0	5	1	200	20	0.4	4.733	0.2598
2	1	−1	0	0	0	15	1	200	20	0.4	5.113	0.4345
3	−1	1	0	0	0	5	2	200	20	0.4	4.869	0.462
4	1	1	0	0	0	15	2	200	20	0.4	3.954	0.2653
5	0	0	−1	−1	0	10	1.5	100	10	0.4	4.960	0.2193
6	0	0	1	−1	0	10	1.5	300	10	0.4	5.649	0.1878
7	0	0	−1	1	0	10	1.5	100	30	0.4	3.661	0.2439
8	0	0	1	1	0	10	1.5	300	30	0.4	3.987	0.3772
9	0	−1	0	0	−1	10	1	200	20	0.2	4.951	0.1722
10	0	1	0	0	−1	10	2	200	20	0.2	4.123	0.2495
11	0	−1	0	0	1	10	1	200	20	0.6	4.190	0.377
12	0	1	0	0	1	10	2	200	20	0.6	2.580	0.098
13	−1	0	−1	0	0	5	1.5	100	20	0.4	4.509	0.1602
14	1	0	−1	0	0	15	1.5	100	20	0.4	5.729	0.3144
15	−1	0	1	0	0	5	1.5	300	20	0.4	4.982	0.5366

(continued)

Table 2 (continued)

Run order	Coded value					Actual setting					Responses	
	A	B	C	D	E	A (A)	B (A)	C (μs)	D (μs)	E (kg/cm^2)	R_a (μm)	OC (mm)
16	1	0	1	0	0	15	1.5	300	20	0.4	4.634	0.1167
17	0	0	0	−1	−1	10	1.5	200	10	0.2	4.475	0.2988
18	0	0	0	1	−1	10	1.5	200	30	0.2	5.201	0.3096
19	0	0	0	−1	1	10	1.5	200	10	0.6	4.034	0.2333
20	0	0	0	1	1	10	1.5	200	30	0.6	3.196	0.4129
21	0	−1	−1	0	0	10	1	100	20	0.4	4.346	0.182
22	0	1	−1	0	0	10	2	100	20	0.4	3.869	0.1335
23	0	−1	1	0	0	10	1	300	20	0.4	4.702	0.4308
24	0	1	1	0	0	10	2	300	20	0.4	4.664	0.2757
25	−1	0	0	−1	0	5	1.5	200	10	0.4	5.677	0.2245
26	1	0	0	−1	0	15	1.5	200	10	0.4	6.131	0.5352
27	−1	0	0	1	0	5	1.5	200	30	0.4	4.524	0.3838
28	1	0	0	1	0	15	1.5	200	30	0.4	3.877	0.2903
29	0	0	−1	0	−1	10	1.5	100	20	0.2	4.574	0.2086
30	0	0	1	0	−1	10	1.5	300	20	0.2	5.446	0.1301
31	0	0	−1	0	1	10	1.5	100	20	0.6	4.093	0.216

(continued)

Table 2 (continued)

Run order	Coded value					Actual setting					Responses	
	A	B	C	D	E	A (A)	B (A)	C (μs)	D (μs)	E (kg/cm^2)	R_a (μm)	OC (mm)
32	0	0	1	0	1	10	1.5	300	20	0.6	4.494	0.3079
33	−1	0	0	0	−1	5	1.5	200	20	0.2	5.466	0.2258
34	1	0	0	0	−1	15	1.5	200	20	0.2	5.641	0.236
35	−1	0	0	0	1	5	1.5	200	20	0.6	3.943	0.2764
36	1	0	0	0	1	15	1.5	200	20	0.6	4.627	0.3576
37	0	−1	0	−1	0	10	1	200	10	0.4	5.531	0.2361
38	0	1	0	1	0	10	2	200	30	0.4	4.113	0.3599
39	0	−1	0	1	0	10	1	200	30	0.4	3.477	0.296
40	0	1	0	1	0	10	2	200	30	0.4	4.468	0.1642
41	0	0	0	0	0	10	1.5	200	20	0.4	4.432	0.276
42	0	0	0	0	0	10	1.5	200	20	0.4	3.833	0.2802
43	0	0	0	0	0	10	1.5	200	20	0.4	4.459	0.4000
44	0	0	0	0	0	10	1.5	200	20	0.4	4.325	0.3107
45	0	0	0	0	0	10	1.5	200	20	0.4	3.890	0.4775
46	0	0	0	0	0	10	1.5	200	20	0.4	4.219	0.4281

Table 3 Chemical composition of work material (Al-SiC MMC)

C	O	Mg	Al	Si	S	Cl	K	Ca	Fe	Cu	Zn
42.75	41.41	0.27	13.0	1.67	0.04	0.14	0.07	0.05	0.08	0.17	0.35

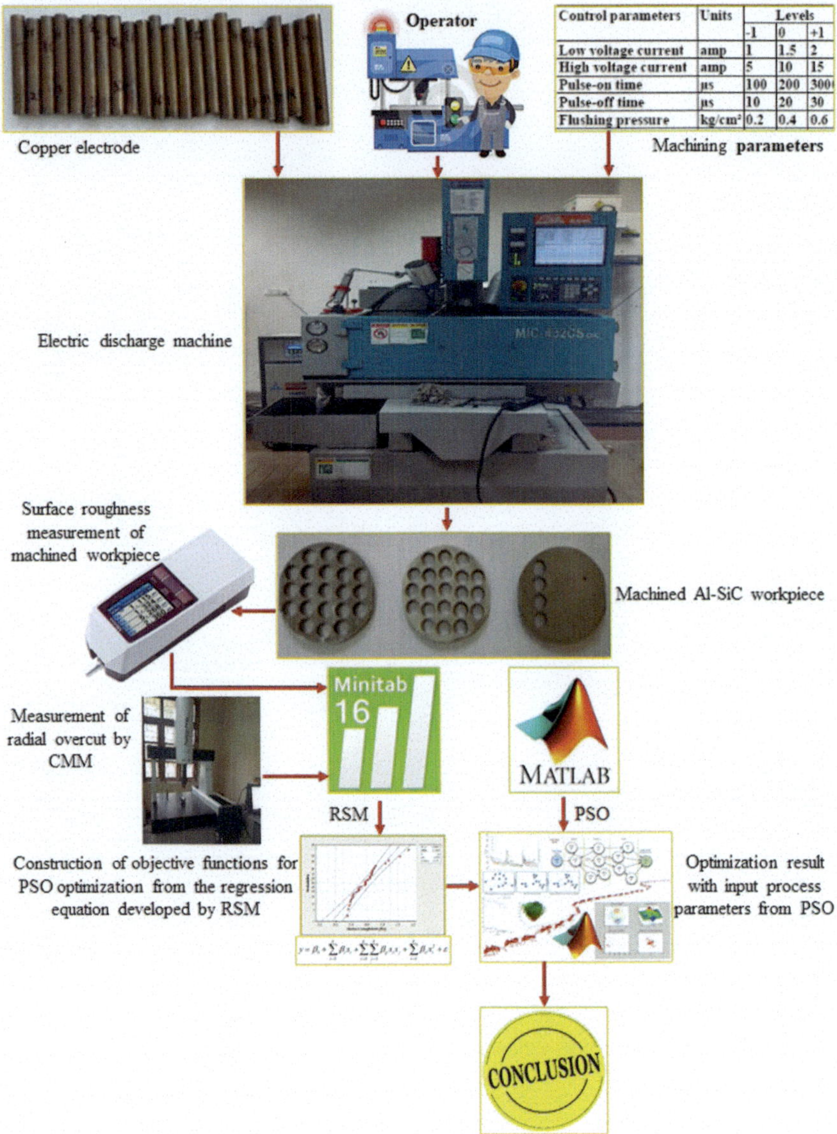

Fig. 1 Layout of experimental setup and methodology presented

3 Results and Discussion

3.1 Model Prediction Using RSM

The results of response characteristics, i.e. surface roughness (Ra) and radial overcut (RO) which were obtained in accordance of BBDOEs, were analysed in Minitab16 software through response surface methodology (RSM), developed the mathematical model to find-out the best fit of correlation between the two responses of the machined component with the pre-cited input machining parameters. Regression equations in the form second order (i.e. quadratic model) for each of the responses (R_a, OC) are presented by;

$$\begin{aligned} R_a = {} & 4.193 + 0.13684A - 0.3209B + 0.17606C - 0.39694D - 0.545E - 0.32375AB \\ & - 0.392AC - 0.57185AD + 0.12725AE + 0.10975BC + 0.76265BD - 0.1955BE \\ & - 0.09075CD - 0.11775CE - 0.391DE + 0.56215A^2 - 0.1271B^2 + 0.31852C^2 \\ & + 0.0472D^2 + 0.04643E^2 \end{aligned}$$

$$R\text{-Sq} = 83.81\%,\ R\text{-Sq (adj)} = 70.31\% \quad (1)$$

$$\begin{aligned} OC = {} & -2.59527 + 0.13952A + 1.44673B + 0.00552C + 0.01864D + 1.88699E - 0.03714AB \\ & - 0.00029AC - 0.00202AD + 0.01775AE - 0.00053BC - 0.00409BD - 0.89075BE \\ & + 0.00004CD + 0.00213CE + 0.02110DE + 0.00036A^2 - 0.19287B^2 - 0.00001C^2 \\ & - 0.00017D^2 - 1.79526E^2 \end{aligned}$$

$$R\text{-Sq} = 70.34\%,\ R\text{-Sq (adj)} = 46.61\% \quad (2)$$

The statistical analysis of developed models for R_a and OC are presented in Table 4 by employing ANOVA in order to check its adequacy and validity. The F value of the model for surface roughness (R_a) and overcut (OC) are 6.21 and 2.96 respectively, which shows the excellent significance of model because of lower magnitude of F table value (2.08). Moreover, it can be clearly noticed that the quadratic models are significant as the P-value is below 0.05. The 'lack-of-fit' P-value of 0.138 for R_a and 0.821 in case of OC (over 0.05), are non-significant concerning the pure error, which is excellent for perfect fitting of the model.

The different surface plots (see, Fig. 2) of surface roughness (R_a) are analysed with the most significant variables, i.e. low voltage current, pulse-off time, flushing pressure. It is noticed that Ra value decreases with the increase of pulse-off time along with flushing pressure and mid-value of low voltage current. Figure 3 illustrates the surface plot of overcut (OC) with low voltage current, high voltage current and pulse-on time, which have been found the most significant parameters. It is noticed that OC decreases with increase of high voltage current but opposite trend occurs with the increase of pulse-on time. It is also observed that OC increases with increasing low voltage current, but it is minimum at high voltage current.

Table 4 ANOVA for surface roughness and overcut

Source	DOF	Seq SS	Adj SS	Adj MS	*F*	*P*	Remarks
(a) Results of ANOVA for surface roughness model							
Regression	20	18.4543	18.4543	0.972271	6.21	0.000	Significant
Linear	5	10.2416	9.9629	1.99258	13.41	0.000	Significant
A (low voltage current)	1	1.2117	1.4857	1.48569	10.00	0.004	Significant
B (high voltage current)	1	0.3162	0.2651	0.26506	1.78	0.194	
C (pulse-on time)	1	0.4960	0.4960	0.49597	3.34	0.080	
D (pulse-off time)	1	3.4654	2.0337	2.03369	13.69	0.001	Significant
E (flushing pressure)	1	4.7524	4.7524	4.75240	31.99	0.000	Significant
Square	5	3.6574	3.8308	0.76615	5.16	0.002	Significant
A^2	1	0.1448	0.1335	0.13346	0.90	0.353	
B^2	1	2.6092	2.5858	2.58545	17.40	0.000	Significant
C^2	1	0.5025	0.8696	0.86962	5.85	0.023	Significant
D^2	1	0.4004	0.0167	0.01671	0.11	0.740	
E^2	1	0.0007	0.0185	0.01848	0.12	0.727	
Interaction	10	4.5553	4.5553	0.4553	3.07	0.012	Significant
A * B	1	0.4193	0.4193	0.41926	2.82	0.106	
A * C	1	0.0482	0.0482	0.04818	0.32	0.574	
A * D	1	1.8465	1.8465	1.84652	12.43	0.002	Significant
A * E	1	0.1529	0.1529	0.15288	1.03	0.321	
B * C	1	0.6147	0.6147	0.61466	4.14	0.053	
B * D	1	0.7091	0.8599	0.85993	5.79	0.024	Significant
B * E	1	0.0648	0.0648	0.06477	0.44	0.515	
C * D	1	0.0329	0.0329	0.03294	0.22	0.642	
C * E	1	0.0555	0.0555	0.05546	0.37	0.547	
D * E	1	0.6115	0.6115	0.61152	4.12	0.054	
Residual error	24	3.5658	3.5658	0.14858			
Lack-of-fit	18	3.1354	3.1354	0.71419	2.43	0.138	
Pure error	6	0.4304	0.4304	0.07173			

(continued)

Table 4 (continued)

Source	DOF	Seq SS	Adj SS	Adj MS	*F*	*P*	Remarks
Total	44	22.0201					
(b) Results of ANOVA for overcut model							
Regression	20	0.375631	0.375631	0.18782	2.96	0.006	Significant
Linear	5	0.059647	0.118106	0.023621	3.73	0.012	Significant
A (low voltage current)	1	0.000027	0.067997	0.67997	10.73	0.003	Significant
B (high voltage current)	1	0.009039	0.067296	0.067296	10.62	0.003	Significant
C (pulse-on time)	1	0.029318	0.042642	0.42642	6.73	0.016	Significant
D (pulse-off time)	1	0.008691	0.005100	0.005100	0.80	0.378	
E (flushing pressure)	1	0.012572	0.019900	0.019900	3.14	0.089	
Square	5	0.099902	0.100204	0.020041	3.16	0.024	Significant
A^2	1	0.018982	0.000698	0.000698	0.11	0.743	
B^2	1	0.004829	0.019221	0.019221	3.03	0.094	
C^2	1	0.032242	0.052177	0.052177	8.23	0.008	Significant
D^2	1	0.000002	0.002350	0.002350	0.37	0.548	
E^2	1	0.043847	0.044728	0.044728	7.06	0.014	Significant
Interaction	10	0.216081	0.216081	0.021608	3.41	0.006	Significant
A * B	1	0.034484	0.034484	0.034484	5.44	0.028	Significant
A * C	1	0.082398	0.082398	0.082398	13.00	0.001	Significant
A * D	1	0.040844	0.040844	0.040844	6.45	0.018	Significant
A * E	1	0.001260	0.001260	0.001260	0.20	0.659	
B * C	1	0.002841	0.002841	0.002841	0.45	0.509	
B * D	1	0.001344	0.001344	0.001344	0.21	0.649	
B * E	1	0.031737	0.031737	0.031737	5.01	0.034	Significant
C * D	1	0.006790	0.006790	0.006790	1.07	0.310	
C * E	1	0.007259	0.007259	0.007259	1.15	0.295	
D * E	1	0.007123	0.007123	0.007123	1.12	0.299	
Residual error	24	0.158406	0.158406	0.006336			
Lack-of-fit	18	0.103384	0.103384	0.005441	0.59	0.821	
Pure error	6	0.055022	0.055022	0.009170			
Total	44	0.534037					

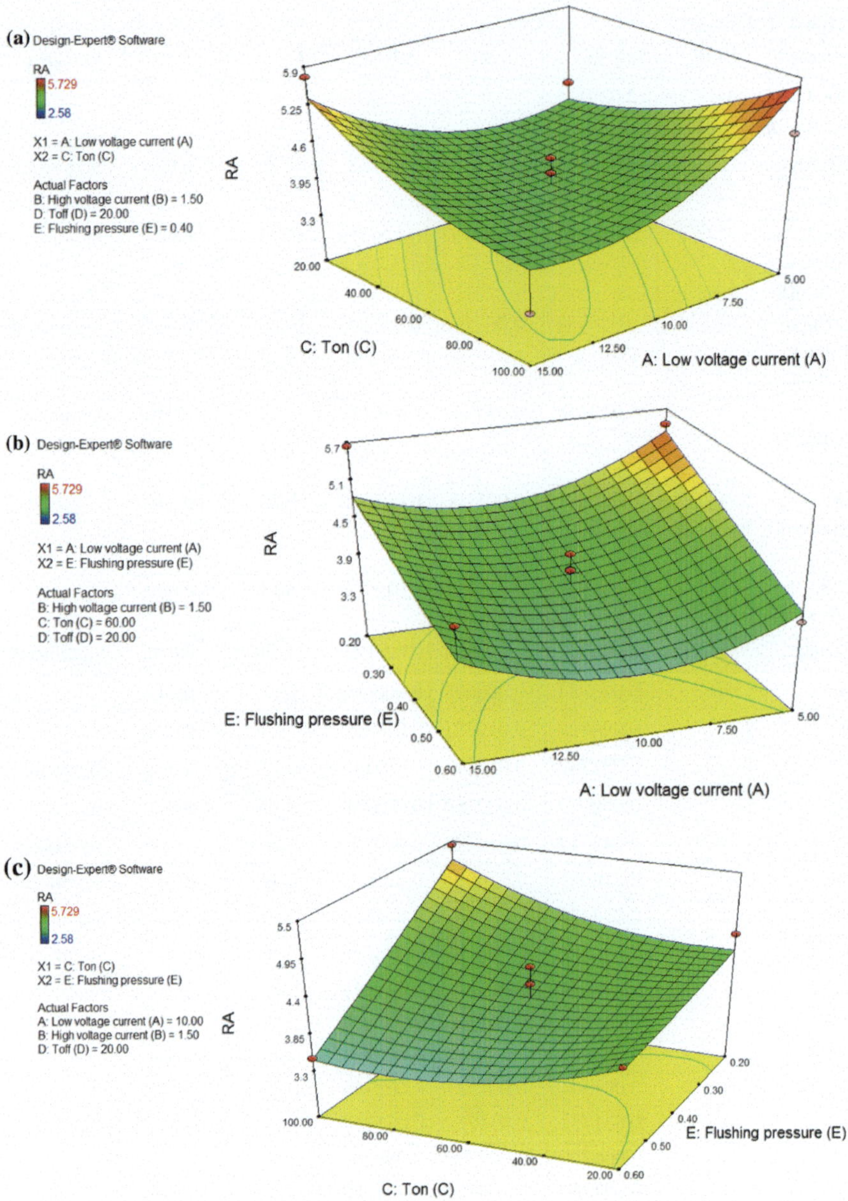

Fig. 2 Parametric influence on surface roughness (R_a)

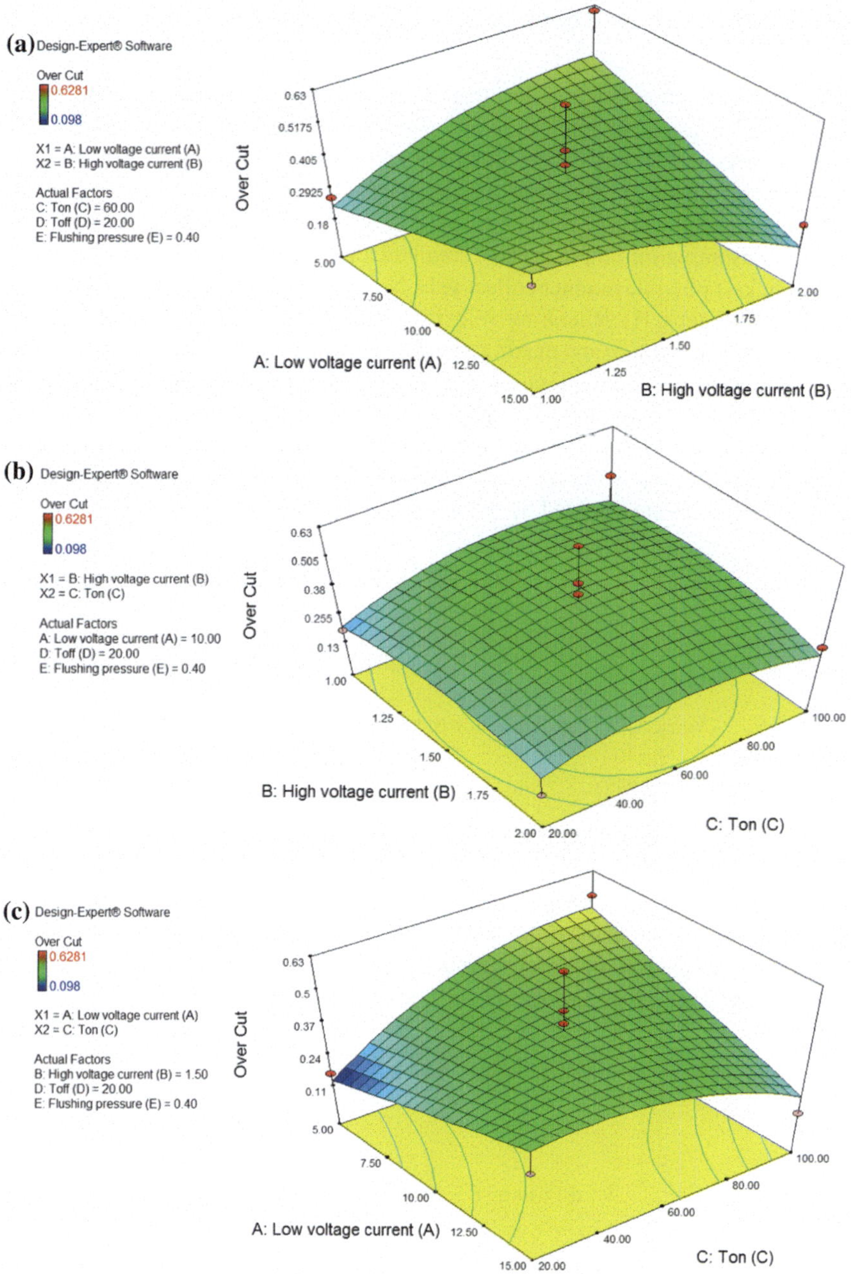

Fig. 3 Parametric influence on overcut (OC)

3.2 Multi-response Optimization Using PSO

Particle swarm optimization (PSO) technique optimizes group activity with inputs from individual group members or particles about respective duration of time to complete the certain problem. This technique is employed as naturally socio-biological colony of organisms like a cluster of birds, fish, swarm of ants, etc. In this case, the particles correct their respective paths in accordance with the path traversed by the particle consuming minimum time to complete the work in order to meet the target in the quickest possible manner collectively. Normally, PSO algorithm involves with the following steps: (i) generating velocity profiles for the assemblage of particles based upon the minimum and maximum value of the input parameters, (ii) correcting velocity of each particle by iteration with respect to its position, (iii) computing the value of fitness function, and (iv) updating the indices $\boldsymbol{p_{best}}$ and $\boldsymbol{g_{best}}$ for the particles signifying individual effort and tem effort, respectively. Meanwhile, if S_{i0} is the starting point and V_{i0} is the speed profile of the ith particle, then the generation of cluster of particles are initiated by the following equations:

$$S_{i0} = S_{\min} + \text{rand}\ (S_{\max} - S_{\min}) \tag{3}$$

$$V_{i0} = \text{Position/Time} = \frac{S_{\min} + \text{rand}(S_{\max} - S_{\min})}{\Delta t} \tag{4}$$

$$\underset{\text{(Velocity of particle } i \text{ at time } t+i)}{V_i(t+i)} + \underset{\text{(Current motion)}}{WV_i(t)} + \underset{\text{(Particle memory influence)}}{\frac{p_{\text{best}_i}(t) - S_i(t)}{\Delta t}} + \underset{\text{(Swarm influence)}}{\frac{g_{\text{best}_i}(t) - S_i(t)}{\Delta t}}, \tag{5}$$

where, S_{max} and S_{min} are the upper and lower limit of input parameters, respectively and Δt is the design space time. With the help of generated particles, the fitness function is used to evaluate the next iteration, i.e. personal best, p_{best} and global best, g_{best}. The updated velocity profile of the particles is given by,

$$S_i(t+1) = S_i(t) + V_i(t+i) \tag{6}$$

and

$$w_i = w_{\max} - \frac{w_{\max} - w_{\min}}{i_{\max}}, \tag{7}$$

where, w_{max} and w_{min} are the maximum and minimum inertia weights, respectively and i_{max} is the maximum number of iterations. It is necessary to evaluate the fitness values of all particles by updating the position and velocity profile of every particle. Such current fitness value is compared with p_{best} and g_{best} values, until termination criteria is met, i.e. representing input parameters as the optimal settings.

Table 5 Deviation of results between the experimental and predicted optimal

Tool and model type		A	B	C	D	E	R_a	OC
Copper	Predicted	10	2	100	30	0.2	4.775	0.139
	Experimental	10	2	100	30	0.2	5.16	0.152
Error							8%	9%

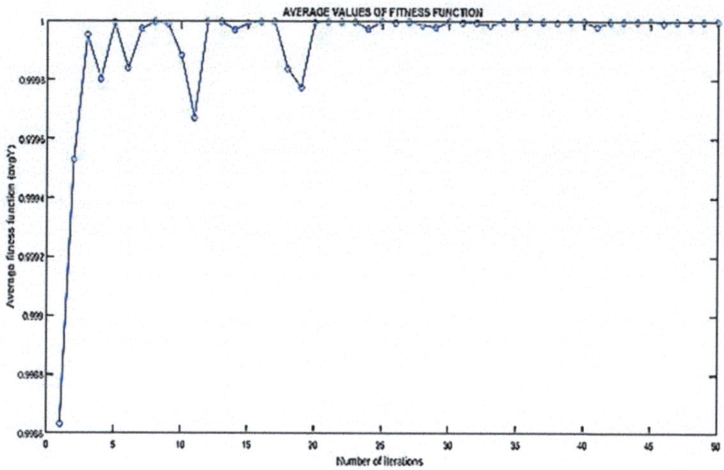

Fig. 4 Convergence curve for PSO algorithm

In the present study, initial swarm size of 50, constants namely cognitive and social scaling parameters (C_1 and C_2) as 2 with maximum number of iterations of 200, and inertia weights of 0.9 and 0.4 for w_{max} and w_{min}, respectively are considered as specific parameters for particle swarm optimization (PSO). The iteration is performed as well as the optimization result for the predicted output is shown in Table 5. It can be seen from Fig. 4 that PSO improves the optimal condition by reducing the number of iterations.

Finally, an additional experiment is performed with the optimal configuration (suggested by PSO) in order to validate the predicted responses at optimal experimental conditions, are listed in Table 5. As can be seen, the experimental result obtained by confirmation test, matches the predicted results obtained by PSO fairly well with a realistic degree of approximation; hence the optimality of the machining condition can be validated. From the analysis of SEM micrograph shows that the machined surface during machining using the predicted optimal parametric combination, shown in Fig. 5.

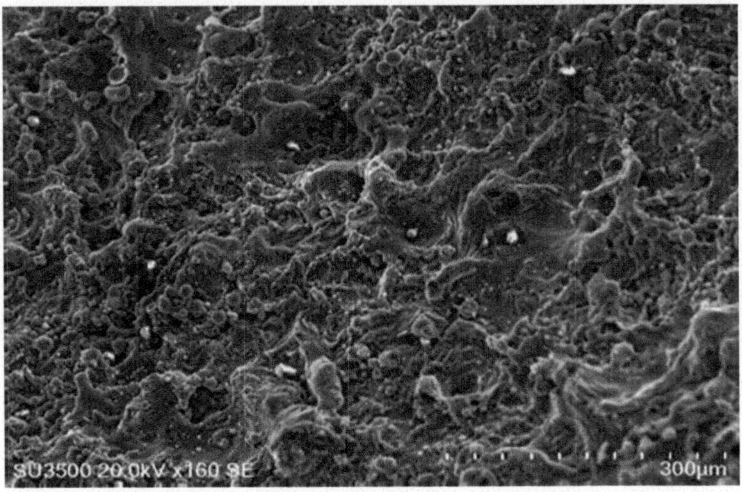

Fig. 5 SEM image illustrating cracks and pores on the machined surface

4 Conclusions

The major objective of the experimental investigation which was to study the effect of pre-cited process parameters on technological response characteristics (R_a and OC) of electrical discharge machining while machining on Al-SiC MMC. The conclusions based on the experimental results are summarized as follows:

- Low voltage current, pulse-off time, and flushing pressure are significant parameters in affecting surface roughness. The Ra decreases steadily by increasing both pulse-off time and flushing pressure.
- The most influential parameters on overcut are low as well as high voltage current, and pulse-on time. The OC can be lowered by applying high voltage current with long pulse durations.
- Empirical models proposed for the technological response characteristics such as surface roughness and overcut have *R*-Square values close to one and *P*-value less than 0.05, which ensured the greater statistical significance with excellence of fit for the model.
- The optimum values of parameters that gives minimum R_a and OC are found to be 15 A (low voltage current), 2 A (high voltage current), 100 μs (pulse-on time), 30 μs (pulse-off time), and 0.2 kg f/mm^2 (flushing pressure).
- The suggested multiple approaches (experimental, statistical, and computational) are reliable methodologies for improving EDM process and can be used in model predictive control, real-time process monitoring, and optimization in different machining process.

References

1. Pramanik A (2014) Developments in the non-traditional machining of particle reinforced metal matrix composites. Int J Mach Tools Manuf 86:44–61
2. Das SR, Kumar A, Dhupal D (2016) Experimental investigation on cutting force and surface roughness in machining of hardened AISI 52100 steel using cBN tool. Int J Mach Mach Mater 18:501–521
3. Das SR, Panda A, Dhupal D (2017) Experimental investigation of surface roughness, flank wear, chip morphology and cost estimation during machining of hardened AISI 4340 steel with coated carbide insert. Mech Adv Mater Mod Process 3:1–14
4. Das SR, Panda A, Dhupal D (2017) Analysis of surface roughness in hard turning with coated ceramic inserts: cutting parameters effects, prediction model, cutting conditions optimization and cost analysis. Ciência e Técnica Vitivinícola: Sci Technol J 32:127–154
5. Panda A, Das SR, Dhupal D (2017) Surface roughness analysis for economical feasibility study of coated ceramic tool in hard turning operation. Process Integr Optim Sustain 1:1–13
6. Das SR, Dhupal D, Kumar A (2015) Study of surface roughness and flank wear in hard turning of AISI 4140 steel with coated ceramic inserts. J Mech Sci Technol 29:4329–4340
7. Padhee S, Nayak N, Panda SK, Dhal PR, Mahapatra SS (2012) Multi-objective parametric optimization of powder mixed electro-discharge machining using response surface methodology and non-dominated sorting genetic algorithm. Sadhana 37:223–240
8. Gopalakannan S, Senthilvelan T, Ranganathan S (2012) Modelling and optimization of EDM process parameters on machining of Al 7075-B_4C MMC using RSM. Procedia Eng 38:685–690
9. Hourmand M, Farahany S, Sarhan AAD, Noordin MY (2015) Investigating the electrical discharge machining (EDM) parameter effects on Al-Mg_2Si metal matrix composite (MMC) for high material removal rate (MRR) and less EWR-RSM approach. Int J Adv Manuf Technol 77:831–838
10. Mohanty S, Mishra A, Nanda BK, Routara BC Multi-objective parametric optimization of nano powder mixed electrical discharge machining of $AlSiC_p$ using response surface methodology and particle swarm optimization. Alex Eng J. https://doi.org/10.1016/j.aej.2017.02.006
11. Sahu J, Mahapatra SS, Puhan D (2013) Multi-response optimisation of electrical discharge machining process using combined approach of RSM and FIS. Int J Prod Qual Manag 12:185–208
12. Surekha B, Swain S, Suleman AJ, Choudhury SD (2017) Performance capabilities of EDM of high carbon high chromium steel with copper and brass electrodes. AIP Conf Proc 1859:1–7
13. Selvarajan L, Manohar M, Kumar AU, Dhinakaran P (2017) Modelling and experimental investigation of process parameters in EDM of SiN_4-TiN composites using GRA-RSM. J Mech Sci Technol 31:111–122
14. Treng CJ, Chen RY (2013) Optimization of electric discharge machining process using the response surface methodology and genetic algorithm approach. Int J Precis Eng Manuf 14:709–717
15. Majumder A, Das PK, Majumder A, Debnath M (2014) An approach to optimize the EDM process parameters using desirability-based multi-objective PSO. J Prod Manuf Res 2:228–240
16. Dey A, Debnath M, Pandey KM (2017) Analysis of effect of machining parameters during electrical discharge machining using Taguchi-based multi-objective PSO. Int J Comput Intell Appl 16:1–16
17. Mohanty CP, Mahapatra SS, Singh MR (2017) An intelligent approach to optimize the EDM process parameters using utility concept and QPSO algorithm. Eng Sci Technol Int J 20:552–562
18. Majumder A (2013) Process parameter optimization during EDM of AISI 316 LN stainless steel by using fuzzy based multi-objective PSO. J Mech Sci Technol 27:2143–2151
19. Rao MS, Venkaiah N (2015) Parametric optimization in machining of Nimonic-263 alloy using RSM and particle swarm optimization. Procedia Mater Sci 10:70–79

Sambeet Kumar Sahu is Assistant Professor in the Department of Production Engineering, Veer Surendra Sai University of Technology (VSSUT), Burla, India. He received his M.Tech. in production engineering from VSSUT, Burla, in 2015.

Subhasree Naik is a PhD student in the Department of Production Engineering, Veer Surendra Sai University of Technology (VSSUT), Burla, India, doing research on electrical discharge machining.

Sudhansu Ranjan Das is Associate Professor in the Department of Production Engineering, Veer Surendra Sai University of Technology (VSSUT), Burla, India. He received his PhD in manufacturing engineering from National Institute of Technology, Jamshedpur, India. His research interests include hard machining, machining process modelling, optimization, analysis and prediction. He has published more than 26 papers in refereed journals and conference proceedings.

Debabrata Dhupal is Professor in the Department of Production Engineering, Veer Surendra Sai University of Technology (VSSUT), Burla, India. He received his PhD in production engineering from Jadavpur University, Kolkata, India. His research areas of interest include laser micromachining, advanced manufacturing process, rapid prototyping and non-traditional machining and metal cutting. He has published more than 65 papers in refereed journals and conference proceedings.

Biogas Plant Slurry Dewatering and Drying Using Hybrid System: A Review

Sujeet Kumar and Om Prakash

Abstract India is the largest milk producing country in the world as well largest animal stock in the world. Wastes generated from animals is the challenge to the society as a whole. In this paper the slurry coming out from biogas plant is used for dewatering and drying has been discussed. The work carried out by many researchers has been collected and discussed. The tables and charts has been drawn from the collected data and comparison has been made. The solid cake after dewatering can be used as a natural fertilizers for our agricultural land and major society problem of biogas plant waste can be solved by suitable use of dewatering and drying technology.

Keywords Wastes · Dewatering · Drying · Belt filter press · Centrifuge

1 Introduction

India is having one of the highest animal stock in the world. It is also the highest milk producing country. The waste of the animals like dungs and urine are the one of the great problem to the society, but if we will see the contents in dungs and urine we realize that it is one of the most valuable products which can be converted so easily as fertilizer for the use of plants which is food of animals and comleting the cycle. Fertilizers is very essential for the plants which has to be supplid to the animals. So the waste from the animal is feed for the plants. Therefore it is challenge to the technocrats and researchers how efficiently we can convert the sludge to valuable products that is Energy and feed to the plants (Fig. 1).

S. Kumar · O. Prakash (✉)
Department of Mechanical Engineering, National Institute of Technology,
Patna 800005, Bihar, India
e-mail: om.prakash@nitp.ac.in

S. Kumar
e-mail: sujeet731@gmail.com

J. Chattopadhyay et al. (eds.), *Renewable Energy and its Innovative Technologies*,
https://doi.org/10.1007/978-981-13-2116-0_10

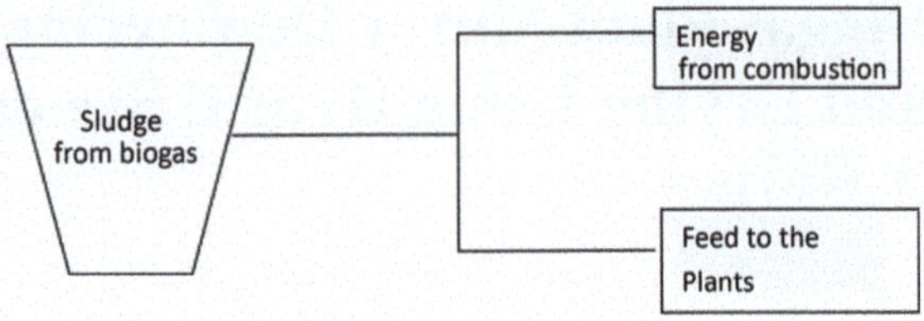

Fig. 1 Diagram for the processing output

Sludges are the mixture of liquid and solid like municipal waste, biogas plant wastes. Dewatering is the process of removing water without evaporation with the use of strainer [1] (Fig. 2).

Due to increase in demand of energy for transportation and other purposes, demand of biogas has also increased [2]. Handling of the biogas residual slurry is very difficult, so slurry has to be treated properly. Biogas residual Slurry after drying (i.e., dewatering) is very useful in many purposes. It can be reused as natural fertilizer and land filling. Sludge treatment and proper disposal are very important for the safety of environment and living creature. The use of solar energy having tremendous amount of energy will reduces the consumption of conventional sources of energy and also improves the drying efficiency of the Biogas Plant slurry dewatering system. With the use of solar dryer, the drying time can be reduced to approx. 65% compared to open sun drying [3]. Slurry from biogas plant is going as a waste due to improper drying technology. But the designed hybrid system can utilise this slurry which will be economical and easy to handle. In market a lot of fertilizer is available which is very costly and also harmful to the plants. Compost slurry is very good alternative to the fertilizers which increase the soil fertility without large expenses. When biogas plant slurry disposed in open environment it increases the pollution in the environment. By drying, pollution can be also controlled. Improper handling of slurry causes water, air and land pollution. Sludge dewatering is used to reduce the moisture content of the sludge (Figs. 3 and 4).

Fig. 2 Belt press dewatering [5]

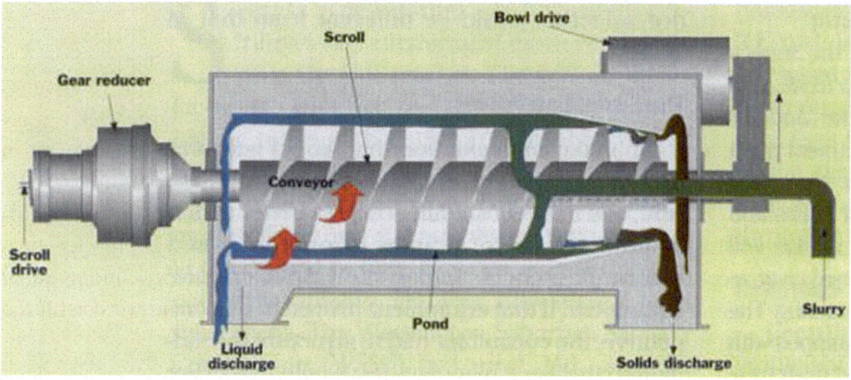

Fig. 3 Centrifuge dewatering [5]

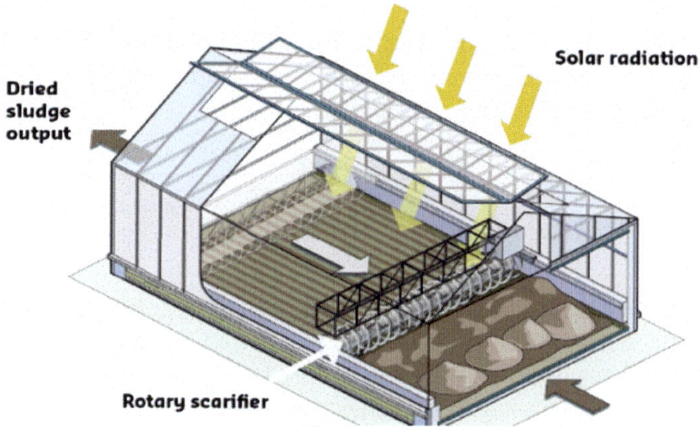

Fig. 4 Solar dryer dewatering [9]

2 Technologies Used for Dewatering and Drying of Biogas Sludge

2.1 Different Dewatering Technologies Used for Sludge Drying

I. Belt Filter Press Dewatering
II. Centrifuge Dewatering

2.1.1 Belt Filter Press Dewatering

Belt filter press consist of two or in some cases three filtering belts that move towards each other forming a closed loop envelope. Sludge is initially dewatered under gravity and then squeezed between the two belts in rollers. Dewatering is achieved in two steps: filtration and pressure squeezing between two rollers. Filtration occurs at the initial gravity drainage zone where no pressure is exerted to sludge. In filtered water, the sludge solids content is not more than 6–10% [4]. Thereafter, progressively increasing the pressure, trapped water within and between sludge particles is some more removed. After both processes finally we can achieve dewatered sludge with solids content between 15% to 25% [4].

2.1.2 Centrifuge Dewatering

Centrifuge Dewatering is used to separate the solid sludge with different densities by spinning action which causes the separation of water from the slurry. Centrifuge Dewatering consumes a large power input [5].

2.2 *Drying Technology*

2.2.1 Solar Sludge Dryer

The solar dryer a newly developed system utilized solar energy to dry the sludge by evaporation. Dewatering using solar energy consume less power with low initial cost but it takes more time to dry as compare to other methods.

3 Findings of Different Researchers

Moss L and P.E. [6] shows the solid capture ability of different dewatering and drying technologies and 92–97 percentage of solid cake remains after drying as shown in Table 1, Chart 1 and Table 2, Chart 2 respectively. Water removed by belt filter press technique is highest up to 87% and by electro dewatering technique only up to 63% of water is removed.

Khoiyangbam et al. [7] explained the Nutrient content of animal fresh excreta before drying and after drying is not diminishing as shown in Table 3, Chart 3 and Table 4, Chart 4 respectively.

Mamais [4] had done comparison between two most dominant technologies in sludge dewatering which are belt filter press and centrifuges dewatering as shown in Table 5. Initial cost and energy consumption in belt filter press is less than centrifuge dewatering, but total cost of centrifuge dewatering is less as compared to belt filter

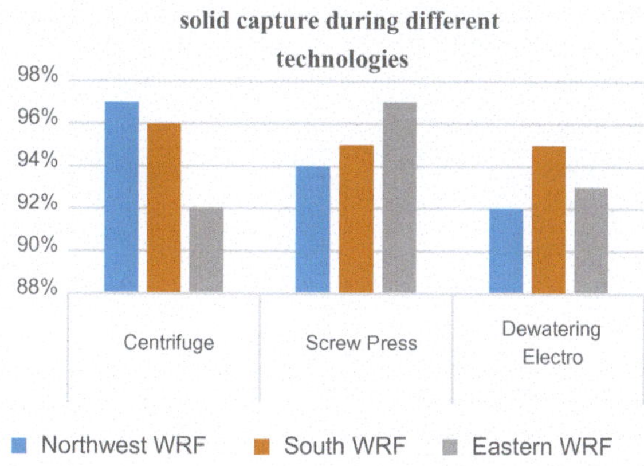

Chart 1 Solid capture in different technologies

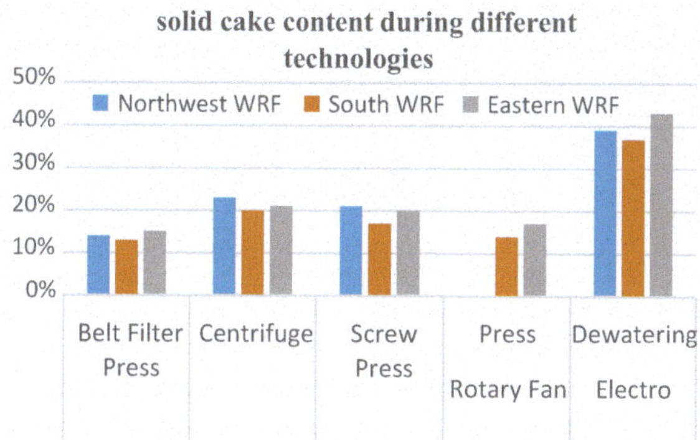

Chart 2 Solid cake content in different technologies

Table 1 Solid capture in different technologies [6]

Percent capture	Centrifuge (%)	Screw press (%)	Electro dewatering (%)
Northwest WRF	97	94	92
South WRF	96	95	95
Eastern WRF	92	97	93

Table 2 Solid cake content in different technologies [6]

Region	Belt Filter Press (%)	Centrifuge (%)	Screw press (%)	Rotary fan press	Electro dewatering (%)
Northwest WRF	14	23	21	–	39
South WRF	13	20	17	14%	37
Eastern WRF	15	21	20	17%	43

Table 3 Nutrient content of animal's fresh excreta and slurry residues before dewatering and drying (% on oven dry basis) [7]

Materials	N	P_2O_5	K_2O	Ca	Mg
Pig dung	2.27	3.1	1.8	0.21	0.54
Cow dung	1.74	1.7	0.6	0.37	0.53
Horse dung	1.07	2.1	3.6	0.26	0.49
Camel dung	1.51	0.35	1.8	0.70	0.69
Poultry excreta	2.17	2.0	4.2	2.28	1.39

Chart 3 Nutrient content of animal's fresh excreta before dewatering

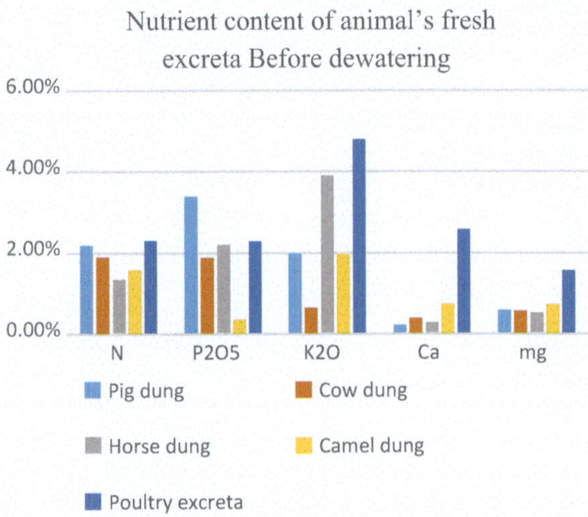

Table 4 Nutrient content of animal's fresh excreta and slurry residues after dewatering and drying (% on oven dry basis) [7]

Materials	N	P_2O_5	K_2O	Ca	Mg
Pig dung	2.20	3.4	2.0	0.23	0.59
Cow dung	1.92	1.9	0.66	0.40	0.58
Horse dung	1.34	2.2	3.9	0.28	0.52
Camel dung	1.60	0.38	2.0	0.76	0.75
Poultry excreta	2.32	2.3	4.8	2.59	1.58

press dewatering because of high use of polymers to separate the additional water from the slurry in case of belt press dewatering. Table 5 comparision of belt press and centrifuges dewatering

Uchiyam [8] explained the variation of pressure and dewatering amount of sludge with time in Belt Press Dewatering technique as shown in Figs. 5 and 6. From these

Chart 4 Nutrient content of animal's excreta after dewatering

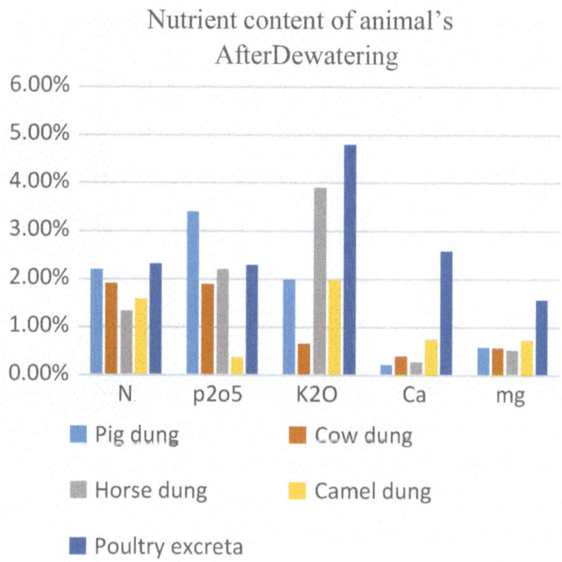

Table 5 Comparision of belt press and centrifuges dewatering

Parameter	Belt press dewatering	Centrifuges dewatering
Moisure content removed	More (83–85%)	Less (77–79%)
Polymer addition	More	Less (25%)
Energy consumption	Less	More
Initial cost	Less	More
Total cost ₹/TON (100 h/week)	8056	6384
Total cost ₹/TON (50 h/week)	8208	6232
Total cost ₹/TON (30 h/week)	11,172	8664

figures one can say that as the dewatering amount of sludge increases, pressure also increases.

4 Result and Discussion

From Chart 1 one can say that solid captured using different technology, centrifuge is the best for Northwest slurry and screw press is best for Eastern whereas for south west screw press and electro are giving the same results.

Chart 2 shows that the solid cake content obtained from slurry by different technique, the belt filter press is best for south, eastern and northwest because it remove more content of water. Press rotary fan is equally good for south slurry and eastern region.

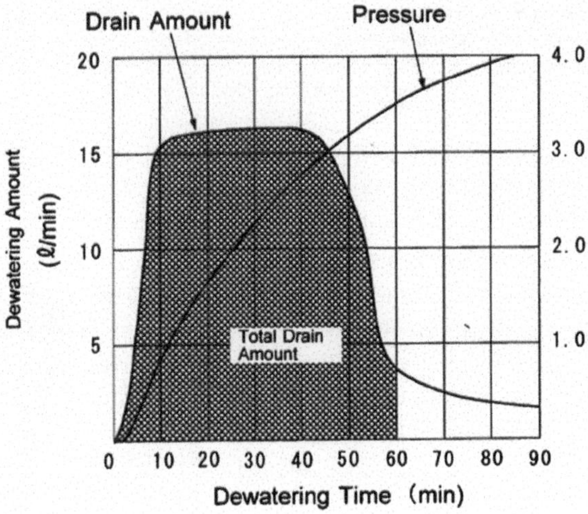

Fig. 5 Dewatering amount versus time [3]

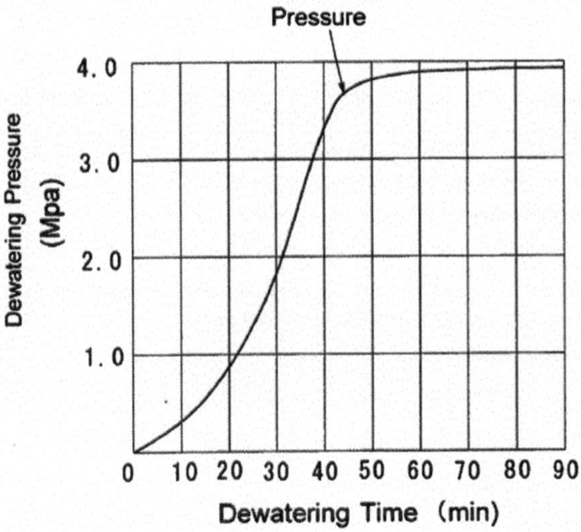

Fig. 6 Dewatering pressure versus time [3]

In Chart 3 the nutrients content of different animal dungs before dewatering and drying has been compared. From the Chart 4 one can say that the poultry excreta contains more potassium oxide (K_2O) and nitrogen (N) whereas the pig dung contains more Phosphorus pentaoxide (P_2O_5),

Chart 4 shows that the Nutrients contains of different animal dungs after dewatering and drying. From this chart one can see that there is no effect on nutrition contents after dewatering and drying.

5 Conclusions

Analyzing the report and data of many researchers at the end one can say that dewatering and drying of slurry will help to solve the problem of society. The solid cake coming out from the dewatering and drying is one of the valuable product to the society.

References

1. Chen G, Lock Yue P, Majumdar AS (2006) Sludge dewatering and drying
2. Ambarish K, Faiz, R (2014) System analysis of de-watering process for treating biogas digestive. Linköping University, Sweden
3. Yuana G, Hongb L, Xua L, Tangc W, Wanga Z (2014) Experimental investigation of a solar dryer system for drying carpet
4. Mamais D, Tzimas A, Efthimiadou A, Kissandrakis J, Andeeadakis A (2009) Evaluation of different sludge mechanical dewatering technologies. J Residuals Sci Technol 6(1)
5. EPA (ed) (2000) Centrifuge thickening and dewatering. United States Environmental Protection Agency, Washington D.C.
6. Moss L and P.E. Trends in dewatering, December, 2012
7. Khoiyangbam RS,Gupta N, Kumar S (2011) Biogas technology towards sustainable development
8. Uchiyam S (2001) Filter press type dewatering. Patent application publication, 30 March 2001
9. Heliantis process diagram natural sludge drying, biosolids (2018) Reduce the quantity of dewatered sludge using solar radiation

Om Prakash is a Professor in the Mechanical Engineering Department, National Institute of Technology, Patna, India. He received his Ph.D. from Indian Institute of Technology, Delhi. His research area of interest is Power Plant, Heat and Mass Transfer, Thermal Engineering and Energy.

Equations of State for Various Dimensional Hard Hyper-sphere Fluids

Sumit Kaur, Binay Prakash Akhouri and Praveen Singh

Abstract We confirm the observations of Luban and Michel [Phys Rev A41:6796 (1990)] in five and Santos et al. [J Chem Phys 120:9113 (2004)] in seven dimensions that the equation of state fits the computer simulation data nearly as well as any other proposed form of equation of state for hard hyper-sphere fluids. We also confirm the observations of Song, Mason, and Stratt [J Phys Chem 93:6916 (1989)] that their theoretical predictions in terms of pair correlation functions in all the dimensions fits the computer simulation data very well.

1 Introduction

1.1 Equations of State for Hard Hyper-Spheres

The study of hard sphere systems with more than three dimensions (hard hyper-spheres) is a very convenient and a general test for theoretical and computational techniques. An impressive study of this can be seen in references [1–12]. If we intend to construct a new equation of state, it is essential to have an accurate knowledge of virial coefficients. These coefficients form the bed-rock of theory of fluids at low and medium densities. A large variety of approaches have been employed for knowing these virial coefficients and constructing equations of state. In some of the approaches, there are a large number of virial coefficients which have been derived and defined in order to get high accuracy [4–12], while there are other approaches that are capable of attaining accuracy by using lesser number of virial coefficients,

S. Kaur (✉)
Department of Physics, Nirmala College, Ranchi 834002, Jharkhand, India
e-mail: sumitbharaj@gmail.com

B. P. Akhouri
Department of Physics, Birsa College, Khunti 835210, Jharkhand, India

P. Singh
Department of Mechanical Engineering, Amity University, Ranchi 834002, India

J. Chattopadhyay et al. (eds.), *Renewable Energy and its Innovative Technologies*,
https://doi.org/10.1007/978-981-13-2116-0_11

but with one of the deepest physical understandings and applications. The virial coefficients are the sum of cluster integrals represented by diagrams [13]. Virial coefficients beyond B_{10} are difficult to calculate as the number of diagrams required for specification show abnormal increase. There are many ways of deducing and drawing inferences from the derived virial coefficients. Some of these assumptions are based on the facts that the higher virial coefficients depend on the lower ones. This inference is drawn on the basis of the assumption that the compressibility factor can be expressed in the form.

$$Z = \frac{P_n(\eta)}{Q_m(\eta)} \tag{1}$$

Here P_n and Q_m are polynomials of order n and m, respectively, whose coefficients are combinations of known virial coefficients. These virial coefficients are also used to fix the polynomial structure for both, the numerator and denominator as in case of Padé approximants [13, 14]. We can expand Eq. (1) into the Taylor series and hence obtain the higher virial coefficients. Also, if we come to know the virial expansions of the compressibility factor in terms of number density, we can calculate the thermodynamic properties of fluids. We can also solve the problem of getting an exact equation of state for any fluid if we are able to determine the coefficients of infinite virial series expansion of the compressibility factor in powers of the density [13], i.e.,

$$Z = \frac{P}{\rho k_B T} = 1 + \sum_{n=2}^{\infty} B_n \rho^{n-1} \tag{2}$$

In Eq. (2), B_n is the virial coefficients of the series expansion, Z is the compressibility factor, P is the pressure, ρ is the number density, k_B is Boltzmann's constant, and T is the absolute temperature. As stated earlier, if the exact values of all the virial coefficients are known, then an accurate equation of state for any fluid can exactly determined. But the irony is that the above equations of state have only a few best known virial coefficients. So in order to make virial expansion converge rapidly, we have a method to propose some approximants which should be based on the fact that the series expansion of the proposed approximants must reproduce a given number of the known virial coefficients. The most widely used and commonly in practice are the Padé approximants. Padé approximants are quotients of polynomials in terms of the density or the so called packing fraction y. The packing fraction is in turn the fraction of space $(\sigma/2)^D$ and is covered by the hard-spheres, i.e., $y = \rho v_D(\sigma/2)$, where, σ is its diameter and v_D is the volume of a D-dimensional sphere of radius $\sigma/2$. If, $v_D(\sigma/2) = \pi^{D/2}/[\Gamma(1 + D/2)](\sigma/2)^D$ and, therefore, y may be expressed as $y = \rho\pi^{D/2}/[\Gamma(1 + D/2)](\sigma/2)^D = \pi^{D/2}/[\Gamma(1 + D/2)](1/2)^D \rho^*$, Here, $\Gamma(x)$ is the gamma function and $\rho^* = \rho\sigma^D$ is the reduced density. By introducing $b_n \equiv B_n/\left(v_D\sigma^D\right)^{n-1} = 2^{(D-1)(n-1)} B_n/B_2^{n-1}$ in Eq. (2), we get Eq. (3) expressed in the form below:

$$Z = 1 + \sum_{n=2}^{\infty} b_n y^{n-1} \tag{3}$$

Now we can get the value of the third virial coefficient ($n = 3$) by the relation [2, 15]

$$\frac{B_3}{B_2^2} = 2\frac{B_{3/4}(D/2 + 1, 1/2)}{B(D/2 + 1/2, 1/2)}, \tag{4}$$

where, $B(a, b) = \Gamma(a)\Gamma(b)/\Gamma(a + b)$ is the beta function and $B_x(a, b)$ is the incomplete beta function given by $B_x(a, b) = \int_0^x t^{a-1}(1 - t)^{b-1}\mathrm{d}t$; $Re(a) > 0$, $Re(b) > 0$ and $0 \leq x \leq 1$. Here, both B_2 and B_3 are positive definite for arbitrary D. From Eq. (4) $B_3/B_2^2 = 5/8$ for $D = 3$, $B_3/B_2^2 = 4/3 - 3\sqrt{3}/2\pi$ for $D = 4$, $B_3/B_2^2 = 53/128$ for $D = 5$, and $B_3/B_7^2 = 11/50$ for $D = 7$.

1.2 *Equation of State of Hard Hyper-Sphere in Arbitrary Dimension Using Pair Correlation Function*

The pair correlation function [13] describes the variation of number density of particles as a function of distance from a reference particle. All particles are assumed to be in a spherical shell dR that surrounds the reference particle. If there are N particles contained in a region of size V (which is the volume in D-dimension) then the average number density ρ is given by N/V. Function $G(R)$ is then used to find the probability of observing a particle at R, when it is given that a particle is placed at the origin. This is given by $\rho G(R)\mathrm{d}R$. We find here that $\rho G(R)$ behaves as a local density because when $G(R)$ is zero, there are no particles in dR whereas it is observed that if $G(R)$ is 1, the number of particles in dR is determined by the average density. Also the structure of pure fluids is determined by the pair correlation function. So the equation of state in D-dimensional form for pair correlation function at contact [13] is

$$Z = 1 - (\beta\rho/2D)\int G(R)R[\partial U(R)/\partial R]\mathrm{d}R \tag{5}$$

In the above equation, dR is the D-dimensional differential volume element. Also this equation has been expressed as the sum of "ideal gas part" and "the part caused by particle interaction". The pair potential of the interacting part (for D-dimensional hyper-sphere) is represented as

$$U(R) = \begin{cases} \infty, & R < \sigma \\ 0, & R \geq \sigma \end{cases} \tag{6}$$

In the derivations of equation of state phenomena, Mayer and his co-workers [13], made use of the following function

$$f(R) = \exp[-\beta U(R)], \tag{7}$$

where, $\beta = 1/k_B T$.

We know that in an ideal system, where there are no interactions, the Mayer f-function vanishes everywhere. Also it is a fact that the stronger the interactions, $U(R)$, the larger are the absolute value of f. The pair correlation function may be expanded in powers of density. These assumptions prove very handy and important in giving the idea of cluster expansion and diagrams too. If we apply chain rule to Eq. (7), we can obtain the relation of the form as given below,

$$\partial f(R)/\partial R = [\partial f(R)/\partial U(R)][\partial U(R)/\partial R] = -\beta \exp(-\beta U(R))[\partial U(R)/\partial R] \tag{8}$$

The Mayer function when considered for hard hyper-spheres is given by,

$$f(R) = \begin{cases} 0, & R < \sigma \\ 1, & R > \sigma \end{cases} \tag{9}$$

Therefore, for hard hyper-spheres $f(R) = -1$ for $R < 1$ and $f(R) = 0$ for $R > 1$, which is the Unit step function or Heaviside step function and it can be defined as the integral of the Dirac delta function and hence as a result its derivative is also the Dirac delta function. We can then replace $\partial f(R)/\partial R$ by $\delta(R - \sigma)$ and rearrange terms to obtain:

$$\partial U/\partial R = -\exp[\beta U(R)]\delta(R - \sigma)/\beta \tag{10}$$

Hyper-spherical coordinate in D-dimensional space are shown in Fig. 1. Here the hyper-sphere with radius R is defined by $x^2+y^2+z^2+\cdots = R^2$. We can also construct the volume $V_D(R)$ by adding infinitely thin spherical shells of radius $0 \leq r \leq R$. Expressing this in the equation form, $V_D(R) = \int_0^R S_D(r)\mathrm{d}r$. The volume and surface area of D-dimensional hard hyper-sphere are $V_D(R) = \pi^{D/2} R^D / \Gamma(1 + D/2)$ and $S_D(R) = D\pi^{D/2} R^{D-1} / \Gamma(1 + D/2)$, respectively.

As per the definition the differential volume element $\mathrm{d}R$ is given by the surface area, S_{D-1} times $\mathrm{d}R$. The surface area of a D-dimensional hyper-sphere is [13]

$$S_D = \frac{D\pi^{D/2} R^{D-1}}{\Gamma(1 + D/2)} \tag{11}$$

Now if we substitute the value of $\partial U(R)/\partial R$ in Eq. (5) from Eq. (10) we get,

$$Z = 1 + \frac{(\rho/2)\pi^{D/2} G(\sigma)\sigma^D}{\Gamma(1 + D/2)} \tag{12}$$

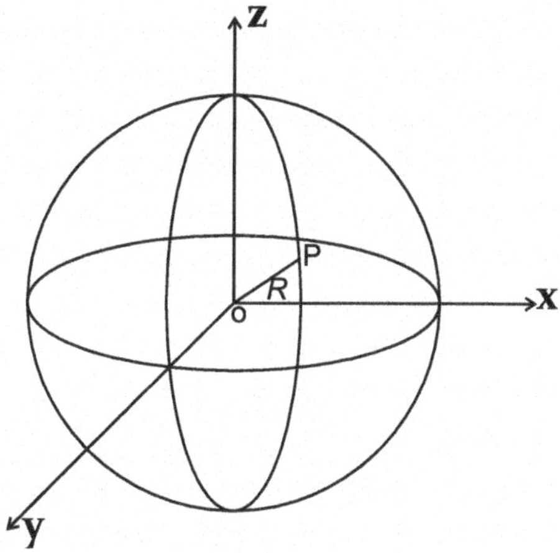

Fig. 1 *D*-dimensional hard hyper-sphere

The second virial coefficient B_2 can be given [4–13] by,

$$B_2 = \frac{\pi^{D/2}\sigma^D}{2\Gamma(1 + D/2)} \tag{13}$$

This leads to an expression for the equation of state in all dimensions as

$$Z = 1 + \rho B_2 G(\sigma) \tag{14}$$

Based on the Carnahan-Starling equation of state [20], a theoretical equation for $G(1)$, using the mean field theory, was derived by Song, Mason and Strat [16] and that can be conveniently written as

$$G(1) = \frac{1 - \alpha y}{(1 - y)^D}, \tag{15}$$

where, y is the packing fraction and $G(1)$ is the pair correlation function for $\sigma = 1$.

$$y = \frac{B_2 \rho^*}{2^{D-1}} \tag{16}$$

and

$$\alpha = D - 2^{D-1}\frac{B_3}{B_2^2}. \tag{17}$$

Here, B_3/B_2^2 can be determined by using Eq. (4). The values of $G(1)$ for five- and seven-dimensional hard hyper-sphere have been calculated by taking different values of packing fraction y and are tabulated in Table 1. The computer simulations as done by many researchers [1–14, 23] for 5-D and 7-D are also summarized in

Table 1. The compressibility factors corresponding to reduced densities for a hard hyper-sphere fluid are given in the same table for comparison of various equations of state defined in text both for 5-*D* and 7-*D*.

2 Aim of the Work

We have a binary aim set up for our work. The first aim is to test the accuracy for a set of equations of state for hard hyper-spheres in 5-*D* and 7-*D*. The second aim is to test the accuracy with the help of an equation for pair correlation function that is based on Carnahan-Starling equation of state. This test for accuracy was done with the computer simulation data. Such methods of Computer simulation data are very common reference tools that are used to test the validity of numerous proposals for the formation of equations of state in *D*-dimensions. The computer simulation data for 5-*D* is on the basis of the work done by Luban and Michel and that for 7-*D* it is based on the work performed by Santos et al.

2.1 Test of Accuracy for a Set of Equations of State in 5-D and 7-D Hard Hyper-sphere

The need of our work is to establish authentic and worthy conclusions about the validity of the analytical expressions that are proposed for the compressibility factor of the hard sphere fluid. So in order to establish the validity we have considered four different equations of state [Colot-Baus [15] (CB), Song-Mason-Stratt [16] (SMS), Luban-Michel [17] (LM), and Maeso-Solana-Amorós-Villar [18, 19] (MSAV)]. In the process of establishing the accuracy of these expressions and also to check their validity for calculating the compressibility factor we had to tally and confirm the results got from the available simulation data given by Luban-Michel in 5-*D* and Miguel Robles, Marinano Lopez de Haro and Andrew Santos [23] in 7-*D*. The exact values of the first four virial coefficients are $B_1 = 1$, $B_2 = 16$, $B_3 = 106$ and $B_4 = 311.18341(2)$ and that of the reduced virial coefficients are $b_2 = 64$, $b_3 = 901.12$ and $b_4 = 4378$.

2.1.1 Colot-Baus Equation of State

Colot and Baus obtained the following two simple expressions for the equation of state by rescaled (truncated) virial expansion:

$$Z = \frac{1 + (B_2 - D)y + [B_3 - B_2 D + D(D-1)/2]y^2}{(1-y)^D} (5-D) \text{ or } (7-D) \tag{18}$$

Table 1 Table represents results in 5-*D* and 7-*D* for: compressibility factor Z from computer simulations and from Eq. (14); pair correlation function $G(1)$ from Eq. (15); packing fraction y from Eq. (16) and the reduced density ρ^*

$\rho*$	$y = \frac{B_2\rho^*}{2^{D-1}}$ for 7-*D*	$y = \frac{B_2\rho^*}{2^{D-1}}$ for 5-*D*	Z_{simul} for 5-*D*	Z_{simul} for 7-*D*	$G(1)$ in 5-*D*	$G(1)$ in 7-*D*	$Z = 1+\rho B_2 G(\sigma)$ in 5-D	$Z = 1+\rho B_2 G(\sigma)$ in 7-D
0.1	0.0037	0.016	–	1.25366	1.087	1.053	1.286	1.248
0.2	0.0074	0.033	1.6530	1.53370	1.183	1.109	1.623	1.524
0.3	0.0111	0.049	–	1.86463	1.289	1.166	2.018	1.826
0.4	0.0148	0.066	2.6240	2.21033	1.408	1.226	2.482	2.159
0.5	0.0185	0.082		2.61742	1.539	1.289	3.025	2.523
0.6	0.0221	0.099	4.0080	3.06504	1.685	1.352	3.661	2.916
0.7	0.0258	0.115	–	3.54495	1.849	1.420	4.406	3.348
0.8	0.0295	0.132	5.9970	4.09897	2.031	1.491	5.276	3.818
0.9	0.0332	0.148	–	4.70135	2.236	1.564	6.296	4.325
1.0	0.0369	0.164	8.7480	5.38910	2.466	1.641	7.490	4.877
1.1	0.0406	0.181	10.523	6.05110	2.725	1.721	8.889	5.472
1.15	0.0420	0.189	11.589	–	2.866	–	9.674	–
1.18	0.0440	0.194	12.217	–	2.955	–	10.177	–
1.2	0.0443	0.197	–	6.81796	3.017	1.804	10.529	6.114
1.3	0.0480	0.214	–	7.63250	3.347	1.900	12.452	6.835
1.4	0.0517	0.230	–	8.51332	3.722	1.981	14.714	7.552
1.5	0.0554	0.247	–	9.42943	4.148	2.075	17.376	8.353
1.6	0.0591	0.263	–	10.4921	4.634	2.173	20.514	9.214
1.7	0.0628	0.280	–	11.5703	5.190	2.275	24.221	10.137
1.8	0.0664	0.296	–	12.6941	5.828	2.378	28.610	11.112
1.9	0.0701	0.313	–	13.9073	6.561	2.489	33.809	12.172
1.95	0.0720	0.321	–	9.03944	6.970	2.547	36.770	12.733

2.1.2 Song-Mason-Stratt Equation of State

An equation of state that was given by Song, Mason-Stratt as

$$Z = 1 + B_2 y \frac{1 + (B_3/B_2 - D)y}{(1 - y)^D} (5-D) \text{ or } (7-D) \tag{19}$$

2.1.3 Luban-Michel Equation of State

This was one of the most accurate proposals that could be given by Luban and Michels for entire densities range of computer simulation data for the fluid phase of hard-spheres and also for four-and five-dimensional hard hyper-spheres. It was a combination of the first four virial coefficients and included two adjustable constants determined from computer simulation data. The analytic expression has the form

$$Z = 1 + \frac{B_2\{1 + [(B_3/B_2) - \eta(y)(B_4/B_3)]y\}}{1 - \eta(y)(B_4/B_3)y + [\eta(y) - 1](B_4/B_2)y^2} (5-D) \text{ or } (7-D) \tag{20}$$

Luban and Michels observed that the computer simulation data for $D = 2 - 7$ shows a linear approximation, $\zeta(y) = p+qy$ to the simulation results for each dimensionality with coefficients that were obtained by a least-square fit;. The values of the adjustable parameters were found to be $p = 1.07416, -5.81$ and $q = 0.35096, 88.2$ in five- and seven dimensions respectively. In Figs. 2 and 3 we have plotted graphs showing variations between the compressibility factors Z in y-direction and the reduced density ρ^* in x-direction.

2.1.4 Maeso-Solana-Amorós-Villar Equation of State

Rescaled Padé approximants proposed by Maeso et al. are given by Eqs. (21) and (22) as

$$Z = \frac{1 + u_1 y + u_2 y^2}{(1 - y)^D (1 + v_1 y)} \quad (D = 5) \tag{21}$$

where $u_1 = (776 - B_4)/36$, $u_2 = (5476 - 11B_4)/36$ and $v_1 = (380 - B_4)/36$.

$$Z = \frac{1}{(1 - y)^D} \frac{B_2 - D + [D(D + 1)/2 + B_2(B_2 - D) - B_3]y}{B_2 - D - [B_3 - B_2 D + D(D - 1)/2]y} (D = 7) \tag{22}$$

Note: While plotting Fig. 3 in seven dimensions the value of reduced virial coefficient,b have been substituted for B in Eqs. (18)–(22). i.e., B should be replaced by b. A sudden drop in compressibility factor is observed at $\rho^* = 1.95$ or $y = 0.0720$ in 7-D. This can be interpreted as the indications of freezing transition.

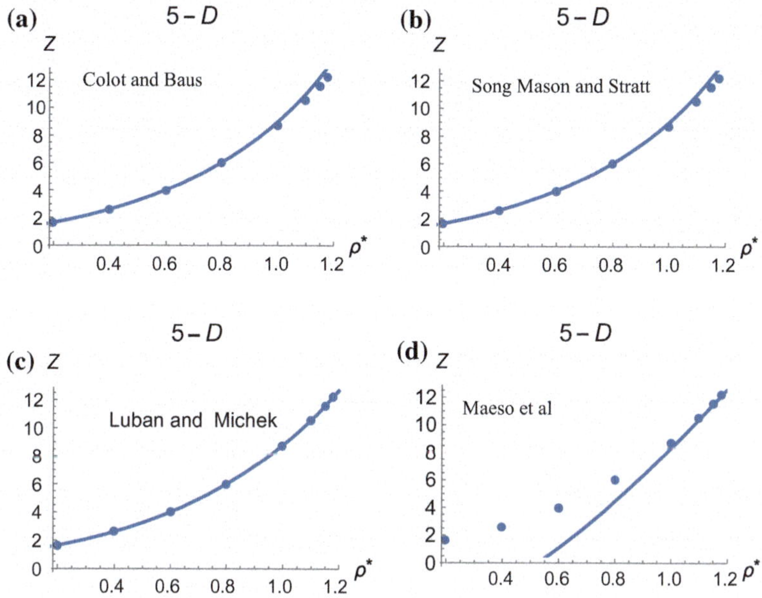

Fig. 2 • Represents the computer simulation data [7, 21, 22] in 5-D

2.2 *Test of Accuracy Using Pair Correlation Function*

A theoretical equation using mean field theory was derived by Song, Mason and Stratt, for $G(1)$ based on the Carnahan-Starling equation of state as shown in Eq. (15). Pair correlation function for various dimensions is calculated and plotted as Fig. 4. This shows the variation of pair correlation function with packing fraction i.e., $1/G(1)$ versus y. The symbols represent the results obtained by computer simulation for various D-dimensional hard hyper-sphere and the solid line represents results obtained from Eq. (15) for various D-dimensions. It can be observed that the theory of Song, Mason and Stratt reproduces the simulation data much closely with the observations.

3 Conclusion

The results of the simulations are used to assess the merits of various proposals that have been made in the literature of equation of state [5–20]. It has been found that the Luban and Michel Equation of state is still one of the most accurate over all density range. The computer simulation data also confirm the theory of Song, Mason and Stratt [16]. We are in excellent agreement with previously published result for the considered dimensions.

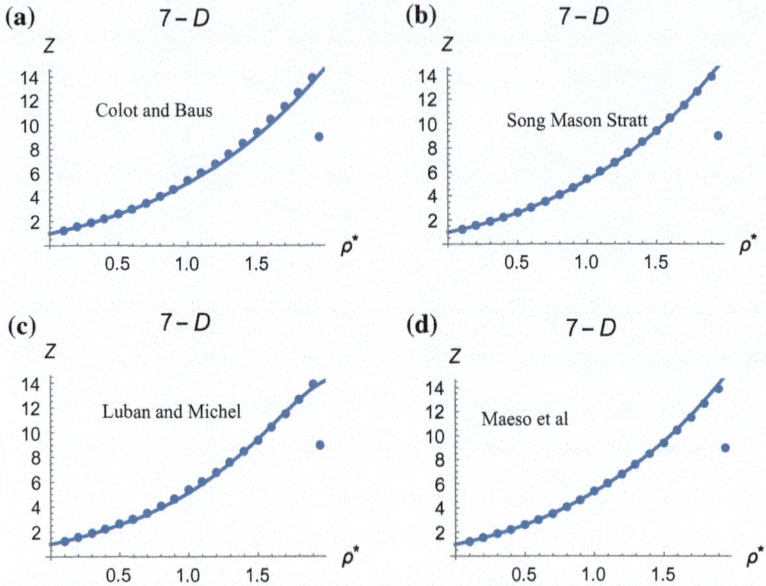

Fig. 3 • Represents the computer simulation data [21, 23] in 7-*D*

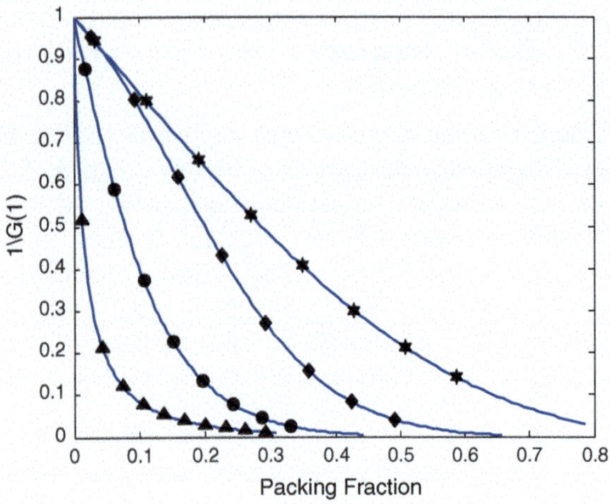

Fig. 4 Song, Maon and Stratt [16] theory in arbitrary dimension. ▲ Computer simulation data in 7-*D*; • computer simulation data in 5-*D*; ◆ computer simulation data in 4-*D*; ✶ computer simulation data in 3-*D*

References

1. Luban M, Bram A (1988) Third and fourth virial coefficients of hard hyperspheres of arbitrary dimensionality. J Chem Phys 6:1976
2. Baus M, Colot JL (1987) Thermodynamics and structure of a fluid of hard rods, disks, spheres, or hyperspheres from rescaled virial expansion. Phys Rev A 36:3912

3. Lyberg I (2008) The fourth virial coefficient of a fluid of hard spheres in odd dimensions. Cond Mat Stat Mech 2:11794–13840
4. Bishop M, Masters A, Vlasov AY (2004) Higher vial coefficients of four and five dimensional hard hyperspheres. J Chem Phys 121(14):6884–6886
5. McCoy BM, Clisby N (2005) New results for virial coefficients of hard spheres in D dimensions. Pramana Ind Acad Sci 64(5):775–783
6. McCoy BM, Clisby N (2004) Analytical calculation of B_4 for hard spheres in even dimensions. J Stat Phys 114:1343–1360
7. Bishop M, Whitlock PA, Klein D (2005) The structure of hyperspherical fluids in various dimensions. J Chem Phys 122(7):074508
8. Bishop M, Andrew MA, Vlasov Yu (2005) The eight virial coefficient of four and five dimensional hard hyperspheres. J Chem Phys 122(15):1882273
9. Lue L, Bishop M, Whitlock PA (2010) The fluid to solid phase transition of hard hyperspheres in four and five dimensions. J Chem Phys 132(10):104509
10. Bishop M, Clisby N, Whitlock PA (2008) The equation of state of hard hyperspheres in nine dimensions for low to moderate densities. J Chem Phys 128(3):034506
11. Bishop M, Whitlock PA (2007) Monte Carlo simulation of hard hyperspheres in six, seven and eight dimensions for low to moderate densities. J Chem Phys 126(2):299–314
12. Whitlock PA, Bishop M, Tiglias JL (2007) Structure factor for hard hyperspheres in higher dimensions. J Chem Phys 126(22):224505
13. Hansen JP, McDonald IR (1986) Theory of simple liquids. Academic press, London
14. Sanchez IC (1994) Virial coefficients and close-packing of hard spheres and disks. J Chem Phys 101:7003
15. Colot JL, Baus M (1986) The freezing of hard disks and hyperspheres. Phys Lett A 119(3):135–139
16. Song Y, Mason EA, Stratt M (1989) Why does the Carnahan-Starling equation work so well? J Che Phys 93(19):6916–6919
17. Luban M, Michels JPJ (1990) Equation of state of hard D-dimensional hypersphere. Phys Rev A 41(12):6796–6804
18. Amros J, Solana JR, Villar E (1989) Equations of state for four and five dimensional hard hypersphere Fluids. Phys Chem Liq 19:119–124
19. Maeso MJ, Solana JR, Amros J, Villar E (1991) Equation of state for D-dimensional hard sphere fluids. Matt Chem Phys 30(11):39–42
20. Carnahan NF, Starling NE (1969) Equation of state for non-attracting Rigid Spheres. J Chem Phys 51:635
21. Rohrmann D, Robles M, de Haro L, Santos A (2008) Virial series for fluids of hard hyperspheres in odd dimensions. J Chem Phys 129:014510
22. Santos A (2008) An equation of state Carnahan-Starling for a five-dimensional fluid of hard hyperspheres. J Stat Mech 8:1–3
23. Robles M, de Haro ML, Santos A (2008) Equation of state of a seven-dimensional hard-sphere fluid. Percus-Yevick theory and molecular dynamics simulations. Cond Mat Stat Mech 129:014510

Sumit Kaur is Assistant Professor in the Department of Physics at Nirmala College in Ranchi University, Ranchi, Jharkhand. She received her M.Sc in year 1995, and the M. Phil in 2012 and presently pursuing Ph.D in Statistical Thermodynamics. She has fifteen years of Senior Secondary and Six years of University teaching experience. She published quite a number of papers and articles. She has interest in the field of Statistical Thermodynamics and non-linear optics.

Binay Prakash Akhouri is Senior assistant Professor in the Department of Physics at Birsa college, Khunti in Ranchi University, Ranchi, Jharkhand. He received his M.Sc (Solid state Physics) in 1992 and obtained his Ph.D in 2004 in Statistical thermodynamics. He has nineteen years of teaching and research experiences. He has published several books and research articles. He has interest for research in the field of statistical thermodynamics and non-linear optics.

Praveen Singh is assistant professor in the Machanical Engineering department, Amity School of Engineering and Technology, Amity University, Ranchi, Jharkhand. Persently pursuing Ph.D at NIT, Jamshedpur.

Energy and Exergy Analysis of Solid Oxide Fuel Cell Integrated with Gas Turbine Cycle—"A Hybrid Cycle"

Tushar Choudhary and Mithilesh Kumar Sahu

Abstract The available waste heat at the gas turbine exhaust has enough potential which can be utilized by integration other thermal systems with the gas turbine cycle. Solid oxide fuel cell (SOFC) is one such promising system that can perfectly been integrated with gas turbine power cycles. In this work, a hybrid power cycle has been developed by integrating recuperative gas turbine cycle with the fuel cell, i.e., SOFC. Through successful thermal integration, the thermal efficiency of conventional gas turbine cycle has enhanced by around 70–80%. This can only be achieved by effective utilization of available waste heat at gas turbine exhaust. The performance of hybrid cycle has been examined from the thermodynamic analysis, i.e., energy analysis and exergy analysis. From the detailed parametric analysis, the influence of turbine inlet temperature, air flow rate, compression ratio, etc. are seen. The thermodynamic performance characteristics of a developed hybrid cycle are compared with the available literature, which shows good agreement. The results obtained from energy and exergy analysis are employed to evaluate the thermodynamic losses in each cycle component and to estimate the work potentials of the fluid streams as well as heat interactions. It has been observed that by increasing turbine inlet temperature (TIT), the performance of hybrid cycle can be increased significantly. In addition to this, the exergy destruction within SOFC decreases as air flow rate increases, whereas in combustion chamber the exergy destruction tends to increase.

Keywords SOFC-GT · Energy · Exergy · Hybrid cycle

T. Choudhary (✉)
School of Mechanical Engineering, VIT Bhopal University,
Kotrikalan, Sehore 466114, India
e-mail: tusharchoudhary311@gmail.com

M. K. Sahu
Mechanical Engineering Department, GVP College of Engineering,
Visakhapatnam 530048, India

J. Chattopadhyay et al. (eds.), *Renewable Energy and its Innovative Technologies*,
https://doi.org/10.1007/978-981-13-2116-0_12

Nomenclature

A	Active surface area (cm^2)
E	Voltage (V)
E_o	Ideal cell voltage at standard conditions (Volts)
ex_{chemical}	Exergy of fuel (chemical) (kJ/kg)
ex_{physical}	Exergy of fuel (physical) (kJ/kg)
ex	Specific exergy (kJ/kg)
Ex	Exergy destruction rate (kW)
F	Faraday constant (C)
h	Enthalpy (kJ/kg)
j	Current density (A/cm^2)
j_o	Exchange current density (mA/cm^2)
S_{gen}	Entropy generation rate (W/K)
LHV	Lower heating value (J/mol)
$\dot{m}$	Mass flow rate (kg/s)
P	Pressure (bar)
$\dot{Q}$	Heat transfer rate (kW)
R	Universal gas constant (J/mol K)
T	Temperature (K)
TIT	Turbine inlet temperature (K)
U_{F}	Fuel utilization ratio
W_{cell}	Power output of cell (W)
$\dot{W}$	Power (kW)

Greek Letter

η	Efficiency
ε	Effectiveness

Subscripts

c	Compressor
dest	Destruction
f	Fuel
gen	Generator
cell, fc	Fuel cell
rc	Recuperator
GT	Gas turbine
PT	Power turbine

1 Introduction

The growing need of energy sustainability, dependency over fossil fuel, avoid the continuing climate are only be accomplished by alternative energy technologies. There is wide range of alternative energy technologies; one such technology is the solid oxide fuel cell (SOFC). This class of fuel cell is the most promising technology which produces a substantial quantity of high-quality heat and energy at high temperature. Therefore, it can be employed more fittingly in hybrid power generation systems, such as gas turbine (GT) based power generation systems. Integrating SOFC with gas turbine cycle leads to efficient power generation along with ultra-high efficiency [1]. This is because of integration of two or additional power cycles in one system, i.e., fuel cell and gas turbine. SOFC is an electrochemical energy conversion device which converts the chemical energy of fuel and air into electrical energy through electrochemical reaction at high temperature. Thus, it is effluent at high temperature which makes it an appropriate candidate for integration with gas turbine cycles [2]. Apart from these, SOFC has various attributes such fuel flexibility, internal reforming, available in various size and shapes, solid state design, high efficiency, nearly zero emission, etc. Thus, it has extensive applications in residential, industries, aerospace, and automotive sector [3–8]. Consequently, SOFC technology has gained considerable interest of researchers across the globe, various designs, integration schemes, computational and experimental works have been reported and are discussed in the literature [9–18].

Choudhary et al. [3] integrate SOFC with cogeneration system for residential purpose. In their work CFD analysis of SOFC model has also carried out using COMSOL. Moreover, parametric analysis has been performed and concludes that concerning to thermo-economics internal reforming is beneficial over the external reforming system.

Moussawi et al. [4] conducted 4 E analyses with multi-objective optimizations technique for SOFC-CCHP system modeled for residential building. They adopted two operation strategies i.e. on-grid base load operations and off-grid following electrical load. Their system under on-grid base load operation yields maximum energy and exergy efficiencies of 65.2 and 45.77% respectively and minimum system cost rate of 22.2 cents/kWh.

Palomba et al. [5] for industrial purpose SOFC based for tri-generation systems is presented for simulating the gasification-SOFC and the whole system ASPEN and TRNSYS has been used. Three different configurations have been presented and the configurations are differing for the cold production and heat back-up system. They conclude that all configurations are energy efficient and have 50% less CO_2 emissions.

For aerospace application, Choudhary et al. [3] integrate SOFC with gas turbine hybrid System for aircraft power generation in unmanned aerial vehicle (UAV). They achieved overall thermal efficiency of 62% which is significantly higher than conventional gas turbine based system. For automotive purpose, Rajashekara et al. [7] presented the comparative study of the SOFC hybrid vehicle strategies for automotive propulsion applications, they discussed the feasibility and challenges occurred during hybridization and compared the performance characteristics of the system. Dimitrova et al. [8] presented an environomic design of SOFC integrated electric vehicles. Their innovative SOFC module has capability to extend the range of vehicle over 600 km and cuts CO_2 emission by environomic optimization.

To maintain even temperature distribution within fuel cell stack, Al-Qattan et al. [9] presented a novel stack design which reduces spatial variation of current and temperature of fuel cell stack. For maintaining the heat generation within the stack, an evenly fuel distribution system has also been presented. Ding et al. [10] fabricated a novel SOFC stack by employing cone shaped tubular design which are connected in series. Their fabricated stack is powered by moist hydrogen and at 800 °C, 1.78 W cm^{-2} of maximum power density has been achieved.

Mahisanan et al. [11] analyzed the SOFC based oxyfuel combustion systems with the recycling of anode and steam. It has found that anode recycling delivers superior performance and has high electrical efficiency and concludes that anode recycling is perfect integration scheme with gas turbine.

Chen et al. [12] presented a novel control strategy for a SOFC-GT system including anode and cathode recirculations. They found that using the developed control scheme the temperature of anode and cathode channel can effective been controlled and helpful in avoiding huge temperature gradient in fuel cell. Worall et al. [18] experimentally utilizes the micro-tubular SOFC waste heat, for dehumidification and cooling purpose. Their proposed tri-generation scheme can be used in building application. Their proposed scheme has electrical efficiency of 11% and regenerator efficiency of around 37%.

In the author previous work [13–15] a novel hybrid cycle has been developed by integration SOFC with cooled gas turbine. Where air film cooling technique has been adopted to cool the gas turbine blades and a unique performance map was plotted. Moreover, 2nd law optimization have also been carried out by using entropy generation minimization (EGM). Apart from this integration, computational analysis [16, 17] of anode supported SOFC has also been carried out using COMSOL. Where, performances of SOFC were examined under steady and transient mode.

The main objective of this work is to perform comparative thermodynamic performance analysis between conventional GT cycle and SOFC-GT hybrid cycle. The energy and exergy analyses performed in this work, using energy and exergy model of each component are used to estimate the performance characteristics of the cycles. Moreover, the impact of operating parameters i.e. compression ratio (rp) and turbine inlet temperature (TIT) on performance parameter such as exergy destruction and thermal efficiency were analyzed.

2 Thermodynamic Model Solid Oxide Fuel Cell-Gas Turbine

In this section mathematical modeling of SOFC-GT hybrid cycle have been presented. For analysis Black box approach was adopted to simulate the mass and energy balance equation of each component of the hybrid cycle. There are 6 components in the hybrid cycle i.e. low-pressure compressor, recuperator, fuel cell (SOFC), after burner, gas turbine, power turbine. The same can be seen in Fig. 1 which shows the schematic of SOFC-GT hybrid cycle and conventional GT cycle.

2.1 Compressor

From 1st and 2nd law the efficiency (1st and 2nd), Power generated, Exergy Destruction for compressor is given as

$$\eta_c = \frac{h_{2s} - h_1}{h_2 - h_1} \tag{1}$$

$$\dot{W}_c = \dot{m}_1 (h_2 - h_1) \tag{2}$$

$$Ex_{dest,c} = T_o S_{gen,c} \tag{3}$$

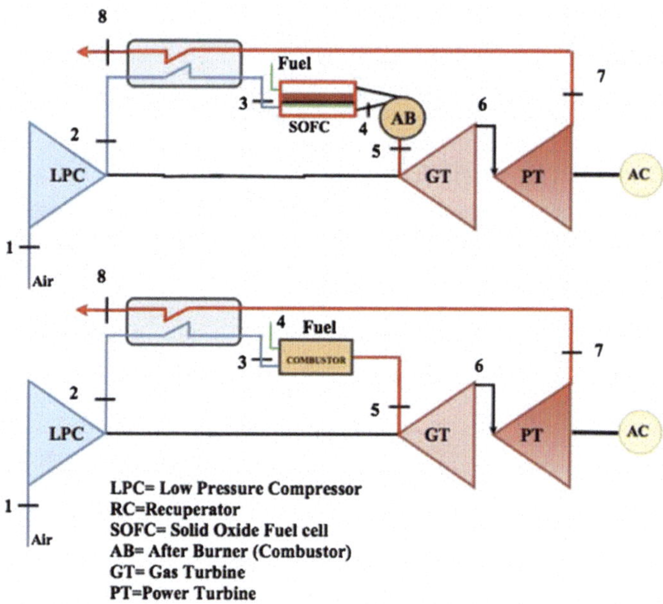

Fig. 1 Schematic of SOFC-GT hybrid cycle and conventional GT cycle

$$\eta_{\text{II,c}} = \frac{\dot{m}_3(ex_2 - ex_1)}{\dot{W}_c} \tag{4}$$

2.2 Recuperator

In recuperator effectiveness, Exergy destruction and efficiency is given as

$$\varepsilon = \frac{T_3 - T_2}{T_7 - T_2} \tag{5}$$

$$\dot{m}_2(h_3 - h_2) = \dot{m}_7(h_7 - h_8) \tag{6}$$

$$Ex_{\text{dest,Recup}} = T_o S_{\text{gen,Recup}} \tag{7}$$

$$\eta_{\text{II,Recup}} = \frac{\dot{m}_3(ex_3 - ex_2)}{\dot{m}_7(ex_7 - ex_8)} \tag{8}$$

2.3 Gas Turbine

From 1st and 2nd law the efficiency (1st and 2nd), Power generated, Exergy Destruction within gas turbine is given as

$$\eta_{\text{GT}} = \frac{h_5 - h_6}{h_5 - h_{6s}} \tag{9}$$

$$\dot{W}_c = \dot{W}_{\text{GT}} \tag{10}$$

$$Ex_{\text{dest,GT}} = T_o S_{\text{gen,GT}} \tag{11}$$

$$\eta_{\text{II,GT}} = \frac{\dot{m}_5(h_5 - h_2)}{\dot{m}_5(ex_5 - ex_6)} \tag{12}$$

2.4 Power Turbine

$$\eta_{\text{PT}} = \frac{h_6 - h_7}{h_6 - h_{7s}} \tag{13}$$

$$\dot{W}_{\text{PT}} = \dot{m}_6(h_6 - h_7) \tag{14}$$

$$Ex_{\text{dest,PT}} = T_o S_{\text{gen,PT}} \tag{15}$$

$$\eta_{\text{II,PT}} = \frac{\dot{W}_{\text{PT}}}{\dot{m}_5(ex_6 - ex_7)} \tag{16}$$

2.5 Solid Oxide Fuel Cell

In fuel cell electrochemical reaction takes place at anode and cathode interface and releases a pair of electron which is been collected through applied load.

$$CH_4 + 2O_2 \rightarrow CO_2 + 2H_2O \tag{17}$$

Using Nernst equation, the electrical potential developed within the fuel cell is calculated by Eq. (18), which constitutes the partial pressures of electrochemical reaction which includes partial pressure reactants and products [19].

$$E = E_o + \frac{RT}{8F}\ln\left[\frac{P_{CH_4}P_{O_2}{}^2}{P_{CO_2}P_{H_2O}{}^2}\right] \tag{18}$$

Within fuel cell, due to presence of irreversibility's decreases the Nernst potential. These irreversibility's are of three types i.e., concentration, activation and ohmic. The details of these losses are examined in [16, 17].

$$V_{\text{cell}} = E - \Delta V_{\text{lossess}} \tag{19}$$

$$\Delta V_{\text{lossess}} = V_{\text{Activation}} + V_{\text{Ohmic}} + V_{\text{Concentration}} \tag{20}$$

$$\dot{W}_{\text{cell,DC}} = V_{\text{cell}}\, j\, A_{\text{cell}} N_{\text{stack}} \tag{21}$$

Applying 1st and 2nd law in the mass balance, exergy destruction and 2nd law efficiency for fuel cell is computed by using Eqs. (22)–(24)

$$\dot{m}_3h_3 + \dot{m}_{\text{fc}} \cdot U_{\text{F}} \cdot \text{LHV} + \dot{m}_{\text{fc}} \cdot (1 - U_{\text{F}}) \cdot \text{LHV} = \dot{W}_{\text{cell,DC}} + \dot{m}_4h_4 \tag{22}$$

$$Ex_{\text{dest,cell}} = \dot{m}_3ex_3 + \dot{m}_{\text{fc}}ex_{\text{physical}} + \dot{m}_{\text{fc}}U_{\text{F}}ex_{\text{chemical}} - \dot{m}_4ex_4 - \dot{W}_{\text{cell,DC}} \tag{23}$$

$$\eta_{\text{II,cell}} = \frac{\dot{W}_{\text{cell,DC}}}{(\dot{m}_{\text{fc}}\,ex_{\text{physical}} + \dot{m}_{\text{fc}}U_{\text{F}}\,ex_{\text{chemical}}) - (\dot{m}_4\,ex_4 - \dot{m}_3\,ex_3)} \tag{24}$$

In the similar way the mass balance, exergy destruction and 2nd law efficiency for combustor is computed by using Eqs. (25)–(28)

$$\dot{m}_5h_5 + \dot{Q}_{\text{losses}} = (\dot{m}_3 + \dot{m}_{\text{fc}}U_{\text{F}})h_4 + \dot{Q}_{\text{comb}} \tag{25}$$

$$\dot{Q}_{\text{losses}} = \text{LHV}(\dot{m}_{\text{comb}}U_{\text{F}} + \dot{m}_{\text{fc}}(1 - U_{\text{F}})) \tag{26}$$

$$\dot{S}_{\text{gen,comb}} = \dot{m}_5s_5 - \dot{m}_4s_4 - (\dot{m}s)_{\text{fuel,comb}} + \frac{\dot{Q}_{\text{losses}}}{T_{\text{surr}}} + \frac{\dot{Q}_{\text{comb}}}{T_{\text{comb}}} \tag{27}$$

$$\begin{aligned} Ex_{\text{dest,cell}} = {} & \left(ex_{\text{physical}} + ex_{\text{chemical}}\right)\dot{m}_{\text{f,comb}} \\ & + \dot{m}_4ex_4 + \dot{m}_{fc}(1 - U_F)ex_{\text{chemical}} \\ & - \dot{m}_5ex_5 - \left(1 - \frac{T_1}{T_{\text{adiab}}}\right)Q_{\text{losses}} \end{aligned} \tag{28}$$

$$\eta_{\text{II,comb}} = \frac{(\dot{m}_5 ex_5 - \dot{m}_4 ex_4)}{\dot{m}_{\text{fc}}(1 - U_{\text{F}}) ex_{\text{chemical}} + \dot{m}_{\text{f,comb}} \left(ex_{\text{chemical}} + ex_{\text{physical}} \right)} \tag{29}$$

Finally, for analysis complete mass balance, exergy destruction, entropy generation and 1st and 2nd law efficiency of complete cycle is given by Eqs. (30)–(36)

$$\dot{m}_4 = \dot{m}_3 + \dot{m}_{\text{fc}} U_{\text{F}} + \dot{m}_{\text{fc}}(1 - U_{\text{F}}) \tag{30}$$

$$(\dot{m}_1 h_1 - \dot{m}_8 h_8) + \left(\dot{Q}_{\text{comb}} - \dot{Q}_{\text{losses}} \right) = \left(\dot{W}_{\text{PT}} + \dot{W}_{\text{cell,DC}} \right) - (\dot{m}_{\text{fc}} U_{\text{F}} \text{LHV}) \tag{31}$$

$$\dot{W}_{\text{cycle}} = \eta_{\text{inv}} \dot{W}_{\text{cell}} + \eta_{\text{gen}} \dot{W}_{\text{PT}} \tag{32}$$

$$\dot{Q}_{\text{comb}} = \dot{Q}_{\text{comb}} + (\dot{m}_{\text{fc}} U_{\text{F}} \text{LHV}) \tag{33}$$

$$\eta_{\text{cycle}} = \frac{\dot{W}_{\text{cycle}}}{\dot{Q}_{\text{total}}} \tag{34}$$

$$Ex_{\text{dest,cycle}} = \dot{m}_1 ex_1 + \dot{m}_{\text{f}} \left(ex_{\text{chemical}} + ex_{\text{physical}} \right) - \dot{m}_8 ex_8 - \dot{W}_{\text{cycle}} \tag{35}$$

$$\eta_{\text{II,cycle}} = \frac{\dot{W}_{\text{cycle}}}{\dot{m}_f \left(\text{ex}_{\text{chemical}} + \text{ex}_{\text{physical}} \right)} \tag{36}$$

3 Methodology

The governing equations of each component of hybrid cycle have been solved in MATLAB. Using energy and entropy balance equations, the temperature and pressure at each inlet and outlet of component have been calculated. The stepwise calculations of output parameters have been detailed in the flow chart Fig. 2 and the operating parameters of hybrid cycle are given in Table 1.

4 Results and Discussion

The thermodynamic performance of developed SOFC-GT hybrid cycle has been evaluated through simulated the thermodynamic model of each components in MATLAB. Separate results have been evaluated for Brayton regenerative cycle and SOFC-GT cycle and comparative analysis have been made on the basis of 1st and 2nd law of thermodynamics and are detained in Table 2 and the validation of the present work has also been done and mentioned in Table 3 and it finds that the obtained result showing agreement with the work of Haseli et al. [20] and Tse et al. [21].

Table 1 Main simulating parameters of hybrid cycle (SOFC-GT)

Simulating parameters	Values
Cycle inlet pressure	101.35 kPa
Ambient temperature	288 K
Inlet temperature GT	1250 K
Adiabatic temperature	250 K
Pressure ratio	4
Compressor efficiency, η_c	81%
Combustor efficiency, η_{comb}	98%
Gas turbine efficiency, η_{GT}	84%
Power turbine efficiency, η_{PT}	89%
Generator output AC	0.95
Utilization factor of air	0.25
Fuel utilization factor	0.85
SOFC stack temperature	1273 K
Current density	300
E_o	1.01 (V)
Cell area	834 (cm^2)
Faraday	96,496 (Columb/kmol)
R	(8.314 kJ/kmol K)
Normal density—j	0.04
Specific resistance—j_0	2×10^{-3}
Recuperator effectiveness	0.8
LHV methane	50,050 (kJ/kg)
Pressure losses recuperator	4%
Pressure losses stack SOFC	4%
Pressure losses combustor	5%

Table 2 Comparison of thermal performance of conventional GT (Brayton) an SOFC-GT cycle

Result parameters	Conventional GT cycle	SOFC-GT cycle
Thermal efficiency of the cycle (%)	30.10	62.16
Power to drive compressor (kW)	716	716
Entropy generation within the cycle (kW/K)	1.97	2.097
Total power generation within the cycle (kW)	488.30	2524

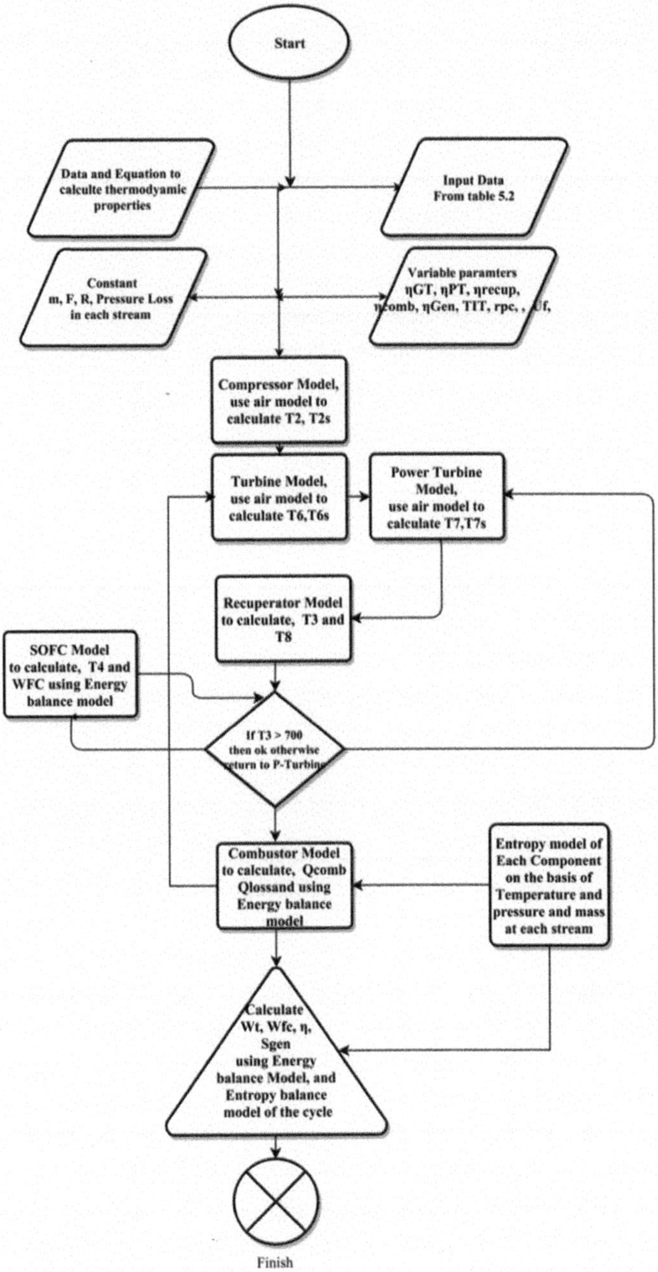

Fig. 2 Process flow chart

Table 3 Validation of present work with the other researcher

Result of parameters	Haseli et al. [20]	Tse et al. [21]	Present
Compression ratio (rp)	4	4	4
Gas turbine inlet temperature (K)	1250	1250	1250
Thermal efficiency of the plant (%)	60.55	59.40	62.15
Exergy efficiency (%)	57.90	–	58.15
Exergy Destruction (kW)	1360.	–	1441.0
Specific power compressor (kJ/kg)	175.00	174.00	173.66
Generator specific power (kJ/kg)	146.00	158.00	172.76
SOFC specific power (kJ/kg)	437.00	440.00	491.94
Net specific power (kJ/kg)	583.90	598.00	600.20
Total power generated (kW)	2419.30	2457.40	2524.00
Air flow rate (kg/s)	4.123	4.110	4.123
Combustor, fuel flow rate (kg/s)	0.0172	0.0177	0.0166
SOFC fuel flow rate (kg/s)	0.0626	0.0645	0.0645

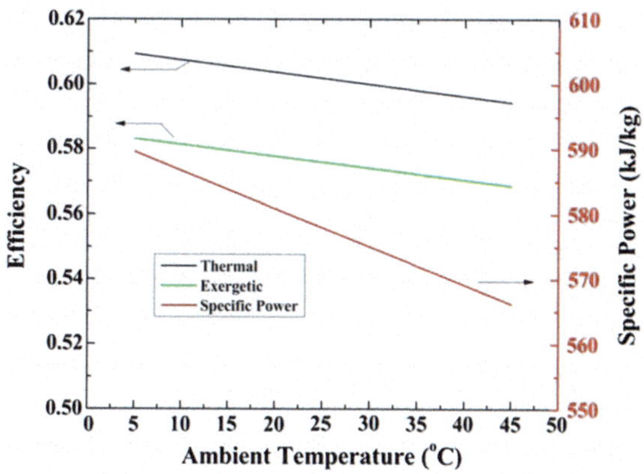

Fig. 3 Effect of ambient temperature on thermodynamic efficiencies and net specific power

4.1 Effect of Ambient Temperature

The impact of ambient condition, mainly temperature on the hybrid cycle has been examined. Figure 3 illustrates the consequence of ambient temperature on specific power and thermal efficiency of hybrid cycle. It has seen the thermal efficiency and exergetic efficiency linearly decreases. This is mainly due to the fact that in colder weather the compressor consumes less power, which results in higher power output and thermal efficiency.

Fig. 4 Variation of turbine inlet temperature on specific power and thermal efficiency of hybrid cycle

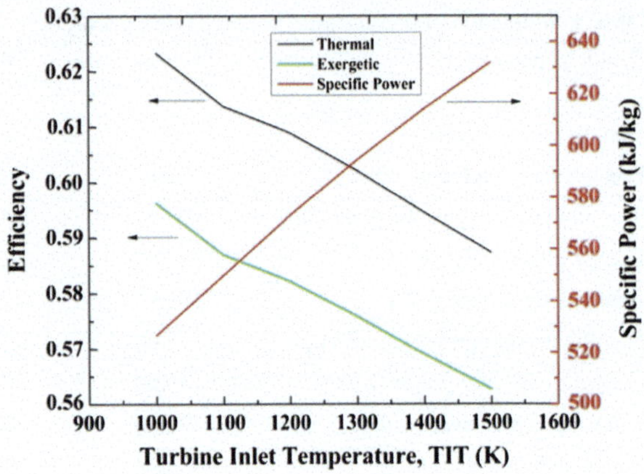

4.2 *Effect of Turbine Inlet Temperature (TIT)*

Figure 4 shows the trend of hybrid cycle specific power and thermal efficiency with respect to variation in turbine inlet temperature. The thermal and exergetic performance of hybrid cycle decreases eventually as TIT increases. Whereas, the opposite trend (increasing) is seen for the specific power as on increasing TIT. From the analysis, it has found that additional heating of hybrid cycle working fluid (gas) subsequently after fuel cell in combustor (afterburner) is not fruitful. Therefore, a combustor is used as after-burner to burn the utilized fuel from the fuel cell. Moreover, the combustor is also useful when the plant is operated under a part-load condition for producing certain power output.

4.3 *Effect of Compression Ratio (rp)*

The comparative performance between conventional GT cycle and SOFC-GT hybrid cycle in terms of energy thermal and exergy efficiency at the same operating conditions is represented in Fig. 5. From the graph, it is seen that at compression ratio 6, the optimal performance has been archived. The performance of hybrid cycle is superior as compared to conventional GT cycle due to having higher power consumption by the compressor is compensated by the fuel cell at higher compression ratio in the hybrid cycle. Whereas, performance concerning to energy (thermal efficiency) and exergy (energy efficiency) the hybrid cycle has 25.6% and 27.8% superior to conventional GT cycle.

Fig. 5 Comparative performance between conventional GT cycle and SOFC-GT hybrid cycle in terms of energy thermal and exergy efficiency

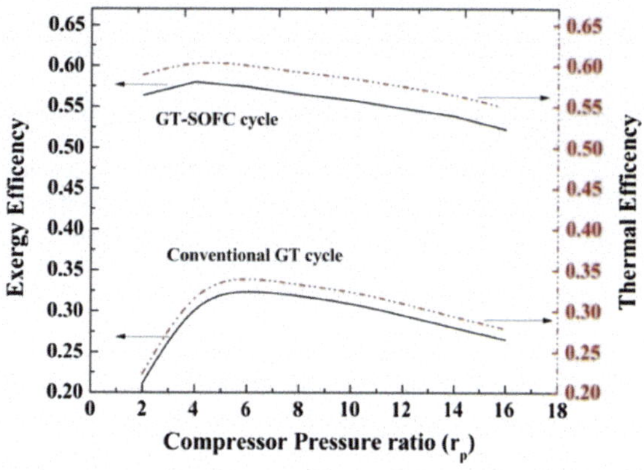

5 Conclusions

In this article thermodynamic modeling of SOFC-GT hybrid cycle has been presented along with energy and exergy analysis. The present work has been validated by comparison with previous work of other researcher Hasel et al. [20] and Tse et al. [21] and the results are within acceptable range. On comparing the performance of proposed hybrid cycle with conventional regenerative Brayton cycle, hybrid cycle yields 71% superior performance.

The optimum performance has attained at 1250 K turbine inlet temperature (TIT) and 6 compression ratio (rp). From the analysis, it can conclude that exergy and thermal efficiency are the strong functions of turbine inlet temperature. As both, the performance parameter reduces as TIT increases. This decrease in thermal and exergy efficiencies is evident from the cycle compression ratio (rp) 6 when the cycle is at optimum (maximum) efficiency. Also, exergy destruction rate of hybrid cycle increases correspondingly as compression ratio and Turbine inlet temperature increases. Moreover, the specific power of the hybrid cycle improves by the increasing TIT, amplifies as TIT increases. However, the performance regarding specific power can be improved by increasing TIT.

References

1. Zhang X, Chan SH, Li G, Ho HK, Feng Z (2010) A review of integration strategies for solid oxide fuel cells. J Power Sources 195(3):685–702
2. Timurkutluk B, Timurkutluk C, Mat MD, Kaplan Y (2016) A review on cell/stack designs for high performance solid oxide fuel cells. Renew Sustain Energy Rev 56:1101–1121
3. Choudhary T, Sahu MK, Sanjay CFD (2017) Modeling of SOFC cogeneration system for building application. Energy Procedia 109:361–368
4. Moussawi HA, Fardoun F, Louahlia H (2017) 4-E based optimal management of a SOFC-CCHP system model for residential applications. Energy Convers Manag 151:607–629

5. Palomba V, Prestipino M, Galvagno A (2017) Tri-generation for industrial applications: development of a simulation model for a gasification-SOFC based system. Int J Hydrogen Energy 42(46):27866–27883
6. Choudhary T, Sahu M, Krishna S (2017) Thermodynamic analysis of solid oxide fuel cell gas turbine hybrid system for aircraft power generation. SAE Technical Paper 2017-01-2062
7. Rajashekara K, MacBain JA, Grieve MJ (2006) Evaluation of SOFC hybrid systems for automotive propulsion applications. IEEE Trans. https://doi.org/10.1109/IAS.2006.256741
8. Dimitrova Z, Maréchal F (2017) Environomic design for electric vehicles with an integrated solid oxide fuel cell (SOFC) unit as a range extender. Renew Energy 112:124–142
9. Al-Qattan AM, Chmielewski DJ, Al-Hallaj S, Selman JR (2004) A novel design for solid oxide fuel cell stacks. Chem Eng Sci 59(1):131–137
10. Ding J, Liu J (2009) A novel design and performance of cone-shaped tubular anode-supported segmented-in-series solid oxide fuel cell stack. J Power Sour 193(2):769–773
11. Mahisanana C, Authayanun S, Patcharavorachot Y, Arpornwichanop A (2017) Design of SOFC based oxyfuel combustion systems with anode recycling and steam recycling options. Energy Convers Manag 151:723–736
12. Chen J, Liang M, Zhang H, Weng S (2017) Study on control strategy for a SOFC-GT hybrid system with anode and cathode recirculation loops. Int J Hydrogen Energy 42(49):29422–29432
13. Choudhary T, Sanjay (2017) Thermodynamic assessment of advanced SOFC-blade cooled gas turbine hybrid cycle. Int J Hydrogen Energy 42(15):10248–10263
14. Choudhary T, Sanjay (2017) Novel and optimal integration of SOFC-ICGT hybrid cycle: energy analysis and entropy generation minimization. Int J Hydrogen Energy, 42(23):15597–15612
15. Choudhary T, Sanjay (2017) Thermodynamic assessment of SOFC-ICGT hybrid cycle: energy analysis and entropy generation minimization Energy 134:1013–1028
16. Choudhary T, Sanjay (2016) Computational analysis of IR-SOFC: thermodynamic, electrochemical process and flow configuration dependency. Int J Hydrogen Energy 41(2):1259–1271
17. Choudhary T, Sanjay (2016) Computational analysis of IR-SOFC: transient, structural integrity, carbon deposition and flow dependency. Int J Hydrogen Energy 41(24):10212–10227
18. Worall M, Elmer T, Riffat S, Wu S, Du S (2017) An experimental investigation of a micro-tubular SOFC membrane-separated liquid desiccant dehumidification and cooling tri-generation system. Appl Therm Eng 120:64–73
19. Choudhary T, Sanjay, Murthy PV (2015) Parametric analysis of syn-gas fueled SOFC with internal reforming. SAE Technical Paper. https://doi.org/10.4271/2015-01-1176
20. Haseli Y, Dincer I, Naterer GF (2008) Thermodynamic modeling of a gas turbine cycle combined with a solid oxide fuel cell. Int J Hydrogen Energy 33(20):5811–5822
21. Tse L, Galinaud F, Martinez-Botas RF. Integration of solid oxide fuel cells into a gas turbine cycle. In: ASME turbo expo power for land. Sea and Air. http://www.felicitas-fuel-cells.info/files/061101_ASME2007_Tse.pdf>2007 (Palais Des Congres Montreal DRAFT)

Dr. Tushar Choudhary is working as Assistant Professor in School of Mechanical Engineering, **VIT Bhopal University**. He has completed his Ph.D. from **National Institute of Technology**, Jamshedpur in 2017. In 2013, he was university topper in Master of Engineering and receive **Gold medal** for academic excellence in his PG by **Dr. K. Radha Krishanan**, Chairman, Indian Space Research Organization. He have 6 years of teaching and research experience, he have successfully handled more than **100 research consultancy projects** for PG and Doctorate candidate's and published various papers in **SCI/Scopus** Indexed international journals. His area of interest is CFD, FEA, Solid oxide fuel cell, Thermodynamic Analysis of hybrid Gas Turbine cycle, Vibration, Turbo machinery.

Mithilesh Kumar Sahu is working as a Assistant professor in the Department of Mechanical Engineering, GVP college of engineering, Visakhapatnam, Andhra Pradesh. He has completed his Ph.D from NIT Jamshedpur in 2018, India doing research on exergoeconomic analysis of air film cooled complex gas turbine cycles and published 6 SCI and 5 SCOPUS paper in international journals. He received his M-Tech. from NIT Jamshedpur India in 2014 in thermal and fluids engineering stream.

Energetic and Exergetic Performance Analysis of a CI Engine Fuelled with Diesel-Blended Plastic Pyrolytic Oil

Amar Kumar Das, Achyut Kumar Panda and Dulari Hansdah

Abstract In this work, plastic oil derived from the kaoline catalyzed pyrolysis of waste polypropylene plastics are used as alternate fuels along with diesel in CI engine. The performance characteristics of the plastic oil blended with 10, 20, 30, 40, and 50% diesel are analyzed and compared with diesel fuel operation. Experiments have been carried out to analyze the Energy and Exergy characteristics of diesel engine fuelled by plastic oil blended diesel fuel. The energetic and exergetic performance of waste plastic oil blended diesel is found higher than the diesel fuel operation.

Graphical Abstract The schematic diagram explains the consumption of air-fuel mixture fed to a diesel engine and extraction of useful work from the engine. It also quantifies different losses of energy in different media associated with the engine during power generation.

A. K. Das
Department of Mechanical Engineering, Veer Surendra Sai University of Technology, Burla, Odisha, India
e-mail: amar.das120@gmail.com

A. K. Panda (✉)
Department of Chemistry, Veer Surendra Sai University of Technology, Burla, Odisha, India
e-mail: achyut.panda@gmail.com

D. Hansdah
Department of Mechanical Engineering, National Institute of Technology Jamshedpur, Jamshedpur, Jharkhand, India
e-mail: dhansdah@gmail.com

J. Chattopadhyay et al. (eds.), *Renewable Energy and its Innovative Technologies*,
https://doi.org/10.1007/978-981-13-2116-0_13

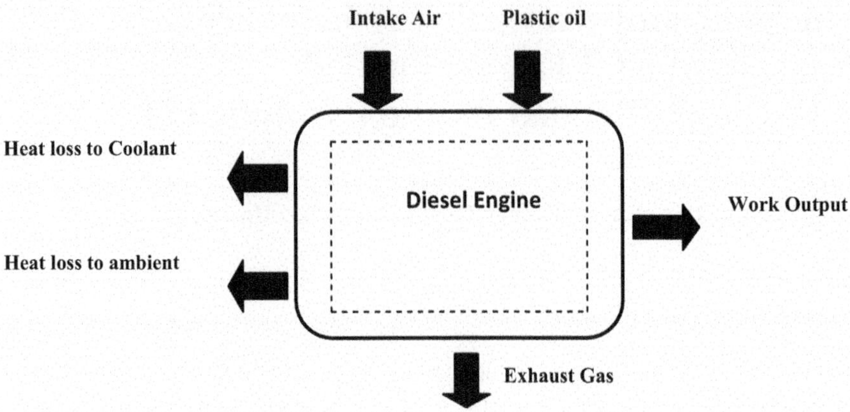

Keywords Catalytic pyrolysis · Waste plastics · Plastic oil · Diesel · CI engine
Energy analysis · Exergy analysis

Nomenclature

m_a	Mass flow rate of air inducted into the combustion chamber under normal ambient conditions (kg/s)
$\dot{m}_f$	Mass flow rate of fuel mixed with air to maintain a required A/F ratio for ease of combustion (kg/s)
W	Shaft work produced in BHP
$\dot{n}_F$	Molar rate of the fuel
h_p and h_r	The enthalpies of the products and reactants per mole of the fuel respectively
ΔH	Enthalpy change due to a change of state at a constant composition
n_{out} and n_{in}	correspond to the relevant coefficients in the reaction equation
H_f^0	Enthalpy of formation
LHV_{fuel}	Lower heating value of the fuel (kJ/kg)
$\dot{E}_{xa}$	Exergy flow rates of inlet air
$\dot{E}_{xf}$	Exergy flow rates of fuel
$\dot{E}_{xg}$	Exergy flow rates of gas
$\dot{E}_{xw}$	Exergy flow rates of shaft work generated
$\dot{E}_{xc}$	Exergy flow rates of heat transfer to coolant and surrounding air
$\dot{E}_{xdes}$	Rate of destructed exergy
$C_{p,a}$	Specific heat capacity of air
P_a	Pressure of the intake air
X_i	Mass fraction
$C_{p,g}$	Specific heat capacity of exhaust gas generated from engine
T_g and P_g	Temperature and pressure of exhaust gas generated from engine

$\dot{W}$	The net work rate
ω	Angular velocity
T	Torque

1 Introduction

The extensive uses of fossil derived fuels are considered unsustainable as such fuel resources are destined to deplete and cause environmental hazards during their uses [1]. Hence, many investigators have been trying to explore various alternate energy resources and improve efficiency of energy consuming device to sustainably meet the growing global energy demands while minimizing the environmental pollutions [2]. On the other hand, the disposal problems of waste plastics and its sustainable utilization is becoming more and more significant in the present scenario. The catalytic pyrolysis of plastic waste to liquid fuel is one of the remarkable ways to recycle the plastic waste and the plastic pyrolysis oil has the matching characteristics and requisition with that of diesel, and thus acts as a substitute of diesel oil in terms of engine performance and energy output [3].

The engine performance in terms of energy output can be understood through energy and exergy analysis. Study on the energetic and exegetic performance of diesel engine under different operational conditions is reported in the literature [4]. Energy analysis done based on the first law of thermodynamics is not sufficient to understand the sustainability of fuel as the limitations of second law of thermodynamics is not considered. However, another quantity exergy being calculated based on the both first and second laws of thermodynamics is closely related to the concept of renewability and sustainability. Exergy analysis is a robust technique for understanding the unsteadiness of a thermodynamic system for energy conversion. Currently, it has become a popular technique to manage with the problems related to the widespread utilization of different energy resources and their subsequent environmental issues [5]. Exergy analysis is also used by researchers to evaluate and optimize the performance of internal combustion (IC) engines running on different (non-renewable and renewable fuels and their blends) fuels [6–9] and extensively reported recently and summarized as follows.

Canakci and Hosoz studied the performance of a four-cylinder turbocharged diesel engine using different biodiesels and petro diesel fuel and compared the performance based on their exergy analysis [10]. The energy and exergy analysis of a diesel engine running with diesel, soybean oil methyl ester, and high-oleic soybean oil was done by Caliskan et al. [11]. Da Costa et al. investigated the performance of a dual-fuel diesel engine running on natural gas and diesel [12]. Exergetic performance assessment of a diesel engine running on oliveepomace oil biodiesel, diesel fuel, and their mixtures was reported by Lopez et al. [13]. Aghbashlo et al. reported the improvement of exergetic performance of a diesel engine using expanded polystyrene as an additive [14]. Caliskan et al. reported an increase in the exergetic efficiency with decreased dead state temperature in a diesel engine fuelled with soybean biodiesel by energy exergy analysis [15]. The performance of a four-cylinder SI engine using three different research octane numbers gasoline fuels was studied by Sayin et al. by energy and exergy analysis [16]. Wiley Sekmen and Yilbasi compared the energetic and exergetic performance of diesel and soybean biodiesel in a four-cylinder, direct injection diesel engine [17]. The effect of compression ratio and injection timing on energy and exergy potential of a palm oil methyl ester run diesel engine was studied by Biplab et al. [18]. A comparative performance of a diesel engine with mahua biodiesel blended diesel by energy and exergy analysis was reported by Panigrahi et al. [19]. Energy and exergy analyses diesel engine for 20% blends of neem oil methylester diesel fuel is carried out by Panigrahi et al. and the blended fuel is found to have low exhaust gas energy loss, higher rate of heat transfer and shows similar trends of energetic and exergetic performance with diesel fuel [20]. From the above studies, it can be concluded that, energy and exergy analysis could help understanding the performance of an alternate fuel, identifying optimum operating condition including fuel blends and engine parameters to design more cost-effective and eco-friendly operations. The objective of the present work is to evaluate the performance of the CI engines fuelled with plastic oil blended diesel.

2 Experimental Investigation

The waste plastic oil used in this experiment is obtained from the kaolin-catalyzed pyrolysis of waste polypropylene at 500 °C with plastics to catalyst ratio 3:1 using a batch reactor reported in our work [3]. The GC-MS composition and physical properties of the plastic oil is summarized in Tables 1 and 2 respectively. The result indicates the presence of different hydrocarbons, mostly alkenes and alkenes along with some fraction of oxygenated organic compounds, such as alcohols and ketones with C_{10}-C_{12}, in the boiling range of 68–346 °C, which infers the presence of a mixture of different oil components equivalent to gasoline, kerosene, and diesel in the oil [3].

The engine performance was carried out on a compression ignition test engine and is well calibrated to reduce the possible errors during experiment. The diagrammatic representation of the test engine setup is shown in Fig. 1 and the research engine specification and parameters details are given in Table 3. The engine is subjected to different conditions and the performance was measured. The load testing has been conducted using a brake drum dynamometer. A required A/F ratio is maintained for better consumption of the fuel. The air box and U-tube manometer is associated as an integral part of air intake system allows excess air to the engine manifold. The mass flow rate of fuel supplied to the engine is measured with the help of a burette and stop watch. The inlet and outlet temperature of water used in the cooling system, the exhaust gas temperature and ambient air temperature was measured using Chomel-Alumel thermocouple displayed with digital temperature indicator. In addition to this, Automatic emission analyser (AVL analyser) and digital smoke meter was used to measure the exhaust emission and Smoke emitted in the engine.

Table 1 GC-MS composition of oil obtained by the catalytic pyrolysis of polypropylene at 500 °C

Peak	Retention time	Molecular formula	Peak	Retention time	Molecular formula
1	3.071	$C_{12}H_{24}$	19	9.144	$C_{13}H_{28}O$
2	3.900	$C_{10}H_{20}$	20	9.264	$C_{13}H_{28}O$
3	4.683	$C_{11}H_{24}$	21	9.382	$C_{11}H_{24}O$
4	4.744	$C_{10}H_{22}$	22	9.761	$C_{12}H_{24}$
5	5.483	$C_{12}H_{24}$	23	11.618	$C_{12}H_{24}$
6	5.543	$C_{12}H_{24}$	24	12.007	$C_{13}H_{28}O$
7	5.771	$C_{12}H_{24}$	25	14.152	$C_{12}H_{24}$
8	5.836	$C_{14}H_{30}O$	26	14.427	$C_{12}H_{24}$
9	6.118	$C_{12}H_{24}$	27	14.503	$C_{12}H_{26}O$
10	6.318	$C_{12}H_{24}$	28	14.527	$C_{12}H_{24}$
11	6.388	$C_{10}H_{20}O$	29	14.878	$C_{12}H_{24}$
12	6.518	$C_{12}H_{24}$	30	16.410	$C_{18}H_{36}$
13	6.617	$C_{13}H_{28}O$	31	18.452	$C_{18}H_{36}$
14	8.261	$C_{10}H_{22}O$	32	18.832	$C_{12}H_{24}$
15	8.739	$C_{12}H_{24}$	33	19.131	$C_{12}H_{24}$
16	8.843	$C_{11}H_{22}$	34	20.312	$C_{18}H_{36}$
17	8.944	$C_{12}H_{22}$	35	22.607	$C_{18}H_{36}$
18	9.081	$C_{11}H_{22}$	36	24.145	$C_{18}H_{36}$

Blended oils (consists of Diesel and 10, 20, 30, 40, and 50% of WPO) were used for testing in the engine. The properties of the blended oil were summarized in the Table 2. These diesel blend waste plastic oil are denoted as 10% BWPO, 20% BWPO, 30% BWPO, 40% BWPO, and 50% BWPO, with the numbers indicating the percentage of WPO in the blend. Higher blends were not used due to detonation in the engine.

The analysis was carried out both quantitatively and qualitatively with an assumption of control volume in the combustion chamber. The experiment was also conducted under constant ambient temperature and atmospheric pressure conditions. The test fuel was subjected to complete combustion assuming no emission loss in the form of C, H_2, CO, H_2O, or free O_2 as the product.

2.1 *Energy Analysis*

The energy analysis is carried out to determine the physical significance of thermodynamics laws to access the amount of energy released during combustion. The concept of control volume attributes the significance of the supplied energy and utilized energy in the system. The analysis is well clarified with the following assump-

Table 2 Physical properties of plastic oil

Properties	Plastic oil	Diesel	10% BWPO	20% BWPO	30% BWPO	40% BWPO	50% BWPO
Density @ 30 °C in (g/cc)	0.7771	0.84	0.83371	0.82742	0.82113	0.81484	0.80855
Boiling point range (°C)	68	150	141.8	133.6	125.4	117.2	109
Gross calorific value (kcal/kg)	11,256	10,162	10,271.4	10,380.8	10,490.2	10,599.6	10,709
Kinematic viscosity, cSt @ 30 °C	2.27	5.3	4.997	4.694	4.391	4.088	3.785
Cetane number	51	55	54.6	54.2	53.8	53.4	53
Flash point (°C)	−12	58	51	44	37	30	23
Fire point (°C)	−12	68	60	52	44	36	28
Pouring point (°C)	−45	–	−4.5	−9	−13.5	−18	−22.5
Cloud point (°C)	−45	–	−4.5	−9	−13.5	−18	−22.5

Table 3 Specification of diesel engine used in the experiment

Engine parameters	Specifications
Make of model	Comet VCT-1
Types of engine	Four-stroke, CI, direct ignition
Cylinder diameter	80 mm
Stroke length	110 mm
Compression ratio (CR)	17.5:1
Rated power	7.4 kW @ 1500 rpm
Engine speed	1500 rpm
Injection pressure	200 bars
Injection timing (CA)	23° BTDC
Cooling type	Water cooled

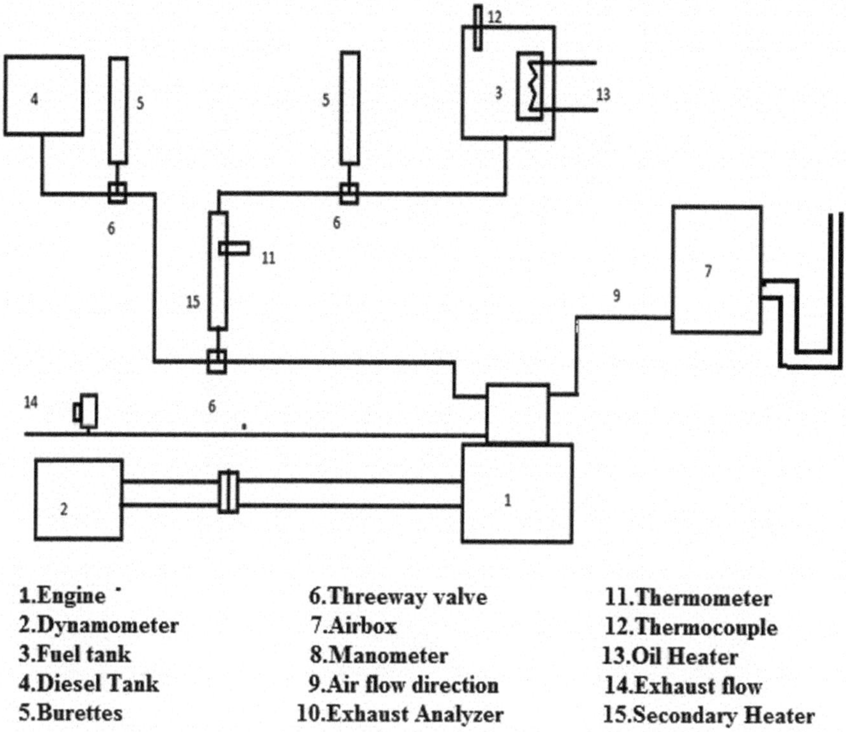

Fig. 1 Schematic representation of the experimental setup

tions; engine operates under steady state conditions, the working medium both in the engine and dynamometer assumed under control volume, the gaseous medium in the engine inlet and outlet considered as an ideal gas mixture, potential energy, and kinetic energy of the fluids throughout the flow is neglected. This analysis gives both the amount of energy present and also discusses about the losses of energy in the system.

Figure 2 explains the mechanism of energy transfer in the form of heat transfer and work transfer in different state of the working fluid. The engine is introduced with a mixture of fuel and ambient air in a required proportion and being ready for combustion. This combustible mixture evolves power which can be encountered as shaft work with some form of energy losses to ambient as exhaust gas emission and heat losses to coolant. The following parameters are taken into consideration for energy calculations.

For any thermodynamic open system [21, 22].

Mass balance equation is given by,

$$\dot{m}_{\text{in}} = \dot{m}_{\text{out}} \tag{1}$$

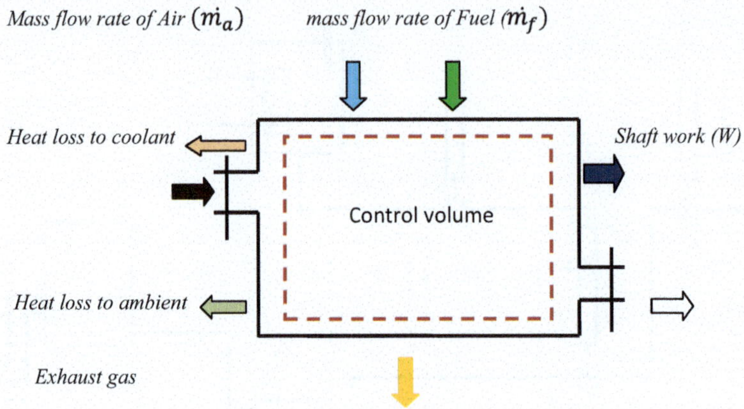

Fig. 2 Schematic diagram of energy flow system in a thermodynamic system

According to steady flow energy equation (SFEE) is given by,

$$\dot{E}_{\text{in}} - \dot{E}_{\text{out}} = \frac{\mathrm{d}E_{cv}}{\mathrm{d}t} \tag{2}$$

$$\dot{Q}_{\text{in}} - \dot{W}_{\text{out}} + \dot{m}_{\text{in}}\left[h_i + \frac{V_i^2}{2} + gZ_i\right] - \dot{m}_{\text{out}}\left[h_e + \frac{V_e^2}{2} + gZ_e\right] = \frac{\mathrm{d}E_{cv}}{\mathrm{d}t} \tag{3}$$

For a steady flow system,

$$\mathrm{d}\dot{E}_{\text{System}} = 0 \tag{4}$$

$$\dot{E}_{\text{in}} = \dot{E}_{\text{out}} \tag{5}$$

$$\frac{\mathrm{d}E_{cv}}{\mathrm{d}t} = 0 \tag{6}$$

Using Eqs. 4, 5 and 6 in Eq. 3 and ignoring the potential energy and kinetic energy being small,

$$\dot{Q}_{\text{in}} - \dot{W}_{\text{out}} = \dot{m}(h_e - h_i) \tag{7}$$

The energy balance per mole of the fuel may be expressed as [23],

$$\frac{\dot{Q}_{\text{in}}}{\dot{n}_{\text{F}}} - \frac{\dot{W}_{\text{out}}}{\dot{n}_{\text{F}}} = \overline{hp} - \overline{hr} = \sum_{\text{product}} n_{\text{out}\left(\overline{H_f^0}+\overline{\Delta H}\right)_{\text{out}}} - \sum_{\text{reactant}} n_{\text{in}\left(\overline{H_f^0}+\overline{\Delta H}\right)_{\text{in}}} \tag{8}$$

where

$$\Delta H = H(T) - H(T_{\text{ref}}) \tag{9}$$

The heat input energy Q_{in} to the diesel engine is the amount of fuel energy content in the supplied fuel and is given by (Eq. 10),

$$Q_{\text{in}} = m_f.\text{LHV} \text{ (kW)} \tag{10}$$

The shaft work can be evaluated by,

$$\dot{Q}_s = \frac{2\pi NT}{6000} \text{ (kW)} \tag{11}$$

Rate of Energy loss in cooling water is given by,

$$\dot{Q}_W = m_w C_{pw}[T_{c2} - T_{c1}] \text{ (kW)} \tag{12}$$

Rate of energy loss in Exhaust gas is given by,

$$\dot{Q}_e = \left[m_a + m_f\right]_{\text{mix}} C_{pe}[T_{e2} - T_{e1}] \text{ (kW)} \tag{13}$$

C_{pe} may be found by equating $\dot{Q}_w$ and $\dot{Q}_e$, and given by

$$C_{pe} = \frac{m_w C_{pw}[T_{c2} - T_{c1}]}{\left[m_a + m_f\right]_{\text{mix}}[T_{e2} - T_{e1}]} \tag{14}$$

Rate of Unaccounted Energy losses is given by,

$$\dot{Q}_u = \dot{Q}_{\text{in}} - \left[\dot{Q}_{\text{in}} - \left\{\dot{Q}_s + \dot{Q}_W + \dot{Q}_e\right\}\right] \tag{15}$$

2.2 *Exergy Analysis*

The performance of an engine can be well evaluated by its qualitative and quantitative approach. The exergy associated with energy is a quantitative assessment of its usefulness or quality. Exergy analysis acknowledges that although energy cannot be created or destroyed, it can be degraded in quality, eventually reaching a state in which it is in complete equilibrium with the surroundings and hence of no further use for performing tasks. The assumptions for exergy analysis are:

- Exergy change of potential, kinetic, electromagnetic and electrostatic are insignificant.
- Steady flow engine system.
- Available energy is from finite energy source.
- The kinetic and potential exergies being small were ignored as compared to total exergy value.

For a direct injection diesel engine, the exergy balance equation can be written as

$$\dot{E}_{xa} + \dot{E}_{xf} = \dot{E}_{xg} + \dot{E}_{xw} + \dot{E}_{xc} + \dot{E}_{xl} + \dot{E}_{xdes} \quad (16)$$

The exergy flow rate of the air to diesel engine can be expressed as follows:

$$E_{x_a} = \dot{m}_a\left[C_{p,a}\left(T_a - T_o - T_o \ln\left(\frac{T_a}{T_o}\right) + RT_o \ln\left(\frac{P_a}{P_o}\right)\right)\right] \quad (17)$$

The specific heat capacity of the intake air can be calculated using the following equation:

$$C_{p_a} = 1.04841 - 0.0003837T_a + \frac{9.45378T_a^2}{10^7} - \frac{5.49031T_a^2}{10^{10}} + \frac{7.92981T_a^2}{10^{14}} \quad (18)$$

Considering the chemical exergy only for determining the energy flow rate of fuel,

$$\dot{E}_{xf} = \dot{m}_f \cdot e_{xf} = \dot{m}_f \cdot \rho \cdot q_{\mathrm{LHV}} \quad (19)$$

The value of chemical exergy factor (ρ) of the fuel and can be calculated by knowing mass fraction of H, C, O, and Sas used in the following equation [21, 24];

$$\rho = 1.0401 + 0.1728\left(\frac{H}{C}\right) + 0.0432\left(\frac{O}{C}\right) + 0.2169\left(\frac{S}{C}\right)\left[1 - 2.0268\left(\frac{H}{C}\right)\right] \quad (20)$$

$$q_{\mathrm{LHV}} = \sum_i X_i q_{\mathrm{LHV},i} \quad (21)$$

The lower heating value of diesel and plastic fuel were considered to be 42,400 and 42,600 kJ/kg respectively Moreover, the chemical formula of diesel and plastic fuel were taken into account as $C_{12}H_{24}$ and $C_{12.66}H_{24.66}O_{0.25}$.

The summation of physical exergy and chemical exergy flow rates can produce exergy flow rate of the exhaust hot gas from engine.

$$\dot{E}_{X_g}^{ph} + \dot{E}_{X_g}^{ch} = \dot{E}_{X_g} \quad (22)$$

The physical exergy flow rate and chemical exergy flow rate of exhaust hot gas can be calculated using the following equation

$$\dot{E}_{X_g}^{ph} = (\dot{m}_a + \dot{m}_f)\left[C_{p,g}\left(T_g - T_o - T_o \ln\left(\frac{T_g}{T_o}\right) + R'T_o \ln\left(\frac{P_g}{P_o}\right)\right)\right] \quad (23)$$

The specific heat capacity of the exhaust hot gas can be specified as [25],

$$C_{p,g} = \sum_i y_i C_{P,i} \quad (24)$$

where y_i and $C_{P,i}$ are the mass fraction and specific heat capacity of each component in the exhaust hot gas from diesel engine.

$$\dot{E}_{X_g}^{ch} = (\dot{m}_a + \dot{m}_f)n\left(\sum_i x_i \in_i + RT_o \sum_i x_i \ln x_i\right) \tag{25}$$

In order to calculate standard molar percentage of different components in the exhaust gas, it is required to calculate the chemical exergy.

The exergy rate of the shaft work generated is equal to the energy rate of net work

$$E_{Xw} = \dot{W} = \dot{\omega}T \tag{26}$$

The exergy loss rate because of heat transfer rate to the cooling water can be computed as:

$$\dot{E}_{X_c} = \dot{m}_w C_{p,w}\left[\left(T_{w,o} - T_{w,\text{in}} - T_o \ln\left(\frac{T_{w,o}}{w,i}\right)\right)\right] \tag{27}$$

The exergy loss rate to the environment as a result of heat transfer can be assessed as follows:

$$\dot{E}_{X_l} = \dot{Q}_L\left(1 - \frac{T_O}{T_L}\right) \tag{28}$$

The heat loss to the environment can be estimated using the energy balance equation (by equating total energy inputs to total energy outputs) as:

$$\begin{aligned}\dot{Q}_l &= \dot{m}_a(h_a - h_o) + \dot{m}_f q_{\text{LHV}} - (\dot{m}_a + \dot{m}_f)(h_g - h_o) \\ &\quad - \dot{m}_w(h_{w,o} - h_{w,i}) - W \text{ Or,} \\ \dot{Q}_l &= m_a C_{p,a}(T_a - T_o) + \dot{m}_f q_{\text{LHV}} - (\dot{m}_a + \dot{m}_f)C_{p,g}(T_g - T_o) \\ &\quad - \dot{m}_w C_{p,w}(T_{w,o} - T_{w,i}) - W \end{aligned} \tag{29}$$

The exergy efficiency (Ψ) of the DI diesel engine can be evaluated as follows [26],

$$\Psi = \frac{\dot{E}_{X_w}}{\dot{E}_{X_a} + \dot{E}_{X_f}} \tag{30}$$

3 Result and Discussion

3.1 Brake Thermal Efficiency

The brake thermal efficiency of an engine depends upon engine loads and blend ratio and their variation in the present experiment as shown in Fig. 3. The thermal efficiency found decreasing with increase in blend ratio at full load conditions. It can be observed from Fig. 3 that the thermal efficiency is 20.83% at full load for diesel. In addition to this, the brake thermal efficiency of the waste plastic oil blend found closer or slightly higher to diesel up to 80% load. This may be due to higher calorific value of WPO–diesel blend than diesel.

3.2 Brake-Specific Fuel Consumption

Brake-specific fuel consumption (BSFC) is a more consistent criterion to signify the performance of an engine for blended fuels due to difference in their density and calorific values. Brake-specific fuel consumption of different WPO–diesel blends with diesel is summarized in Fig. 4.

It can be observed from Fig. 4 that the increase in load results the decrease in BSFC for both diesel and various blended fuels. In addition to this, the BSFC is found decreasing within the variation of an increase of concentration of WPO in

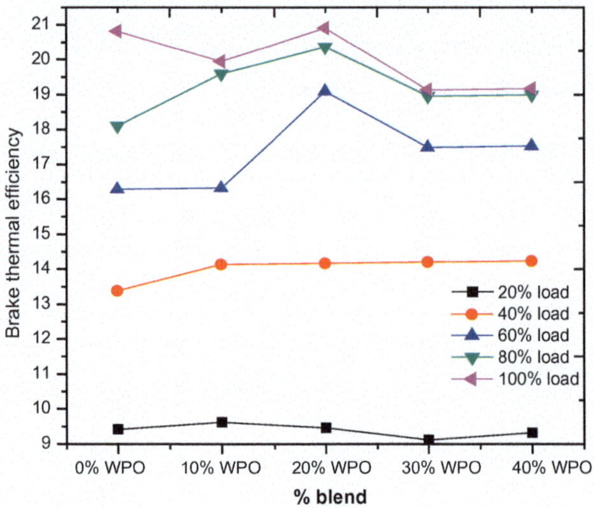

Fig. 3 Brake thermal efficiency of different fuels at different loads

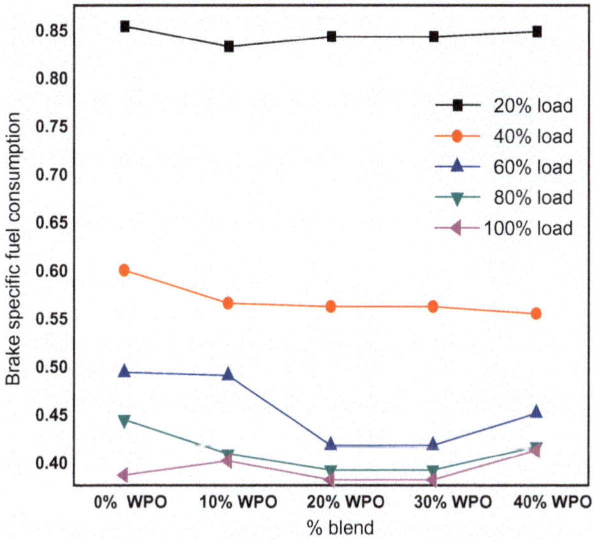

Fig. 4 Brake specific fuel consumption of different fuels at different loads

WPO–Diesel blend. The high calorific value of WPO causes lesser consumption of blended fuel by the engine.

3.3 Energy Analysis of Tested Fuels

The useful energy and energy losses of WPO–diesel blend in comparison with diesel is shown in Fig. 5. The fuel energy generated by WPO–diesel blend is observed very close to diesel at 20% BWPO. The shaft work produced in diesel is found more than WPO. This may be due to higher thermal efficiency of diesel than WPO blend. The heat lost in the cooling water for both fuels are found very close to each other. Energy lost in the form of exhaust emission is found higher in diesel than WPO–diesel blend. This may be due to higher calorific value of diesel than WPO–diesel blend at lower proportion.

3.4 Exergy Analysis of Tested Fuels

The different values of exergy of different fuels are calculated (Eq. 30) and the analysis reveals that the fuel exergy are becoming higher than the corresponding fuel energy inputs.

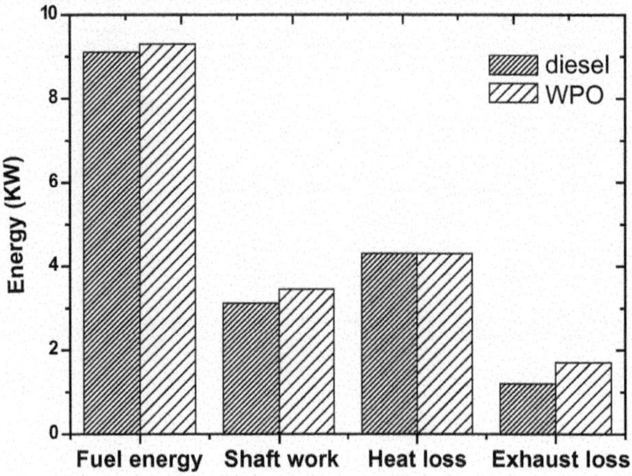

Fig. 5 Energy distribution of WPO in comparisons with diesel

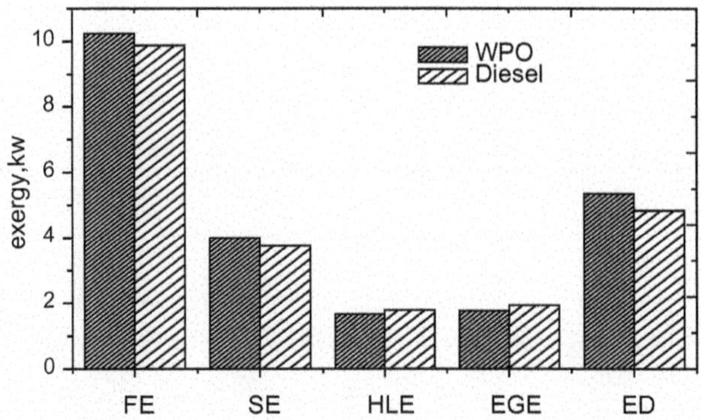

Fig. 6 Exergy analysis of WPO in comparison of diesel. Where FE—fuel exergy, SE—shaft exergy, HLE—exergy due to heat loss to water, EGE—exhaust gas exergy, ED—exergy of destruction

Due to lower specific heating value of diesel, the fuel exergy of diesel is found lower than that of WPO–diesel blend as explained in Fig. 6. The trend is continued for shaft exergy as the brake thermal efficiency of diesel is higher than the blend. In addition, due to higher calorific value of WPO, the exhaust gas exergy is less than diesel.

3.5 Energetic and Exergetic Efficiency of Tested Fuels

Energetic efficiency of the tested fuels agrees at a point and it is resembled with the brake thermal efficiency as shown in Fig. 7. It has been observed that exergetic

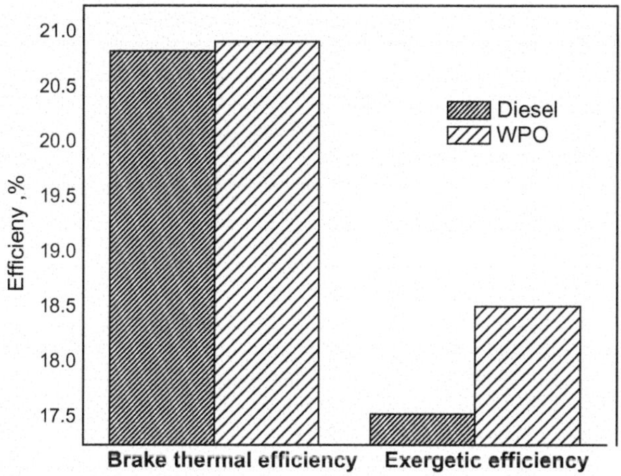

Fig. 7 Comparison of energetic and exergetic efficiency

efficiency of WPO is found more than that of diesel. This may be due to higher BSFC of WPO–diesel blend in comparison to diesel.

4 Conclusion

From the study of conducting performance test using waste plastic oil and diesel on a DI diesel engine, it has been observed that the engine was able to run with maximum 50% of waste plastic oil–diesel blends. The engine is found deteriorating its performance beyond this blend ratio due to knocking and showed better performance up to 30% blend. Brake thermal efficiency of blend is found to be almost the same or marginally higher than diesel up to 80% load and somewhat lower at full load. Exhaust gas temperature is found marginally higher with WPO–diesel blend than diesel operation. Brake specific fuel consumption of WPO is found marginally higher than that of diesel fuel for same rated power. The same trend is followed for destructive exergy of the oil.

References

1. Hasheminejad M, Tabatabaei M, Mansourpanah Y, Javani A (2016) Effect of an emission reducing soluble hybrid nanocatalyst in diesel/biodiesel blends on exergetic performance of a DI diesel 93(2016):353–368
2. Voloshin RA, Kreslavski VD, Zharmukhamedov SK, Bedbenov VS, Ramakrishna S, Allakhverdiev SI (2015) Photo electrochemical cells based on photosynthetic systems: a review. Bio fuel Res 227–235

3. Panda AK, Singh RK (2010) Catalytic performances of kaolin and silica alumina in the thermal degradation of polypropylene. J Fuel Chem Technol 39(3):198–202
4. Mojarrab M, Aghbashlo M, Mobli H (2016) Exegetic performance assessment of a long-life milk processing plant: a comprehensive survey. J Clean Prod 140:590–607
5. Rosen MA, Dincer I (2001) Exergy as the confluence of energy, environment and sustainable development. Exergy Int J 1(1):3e13
6. VanGerpen JH, Shapiro HN (1990) Second-law analysis of diesel engine combustion. J Eng Gas Turb Power 112(1):129–137
7. Rakopoulos CD, Kyritsis DC (2001) Comparative second-law analysis of internal combustion engine operation for methane, methanol, and dodecane fuels. Energy 26(7):705–722
8. Rakopoulos CD, Giakoumis EG (2006) Comparative first-and second-law parametric study of transient diesel engine operation. Energy 31(12):1927e1942
9. Zheng J, Caton JA (2012) Second law analysis of low temperature combustion diesel engine: effect of injection timing and exhaust gas recirculation. Energy 38(1):78–84
10. Canakci M, Hosoz M (2006) Energy and exergy analyses of a diesel engine fuelled with various biodiesels. Energy Source Part B 1(4):379–394
11. Caliskan H, Tat ME, Hepbasli A, Van Gerpen JH (2010) Exergy analysis of engines fuelled with biodiesel from high oleic soybeans based on experimental values. Int J Exergy 7(1):20–36
12. da Costa YJR, de Lima AGB, Bezerra Filho CR, de Araujo Lima L (2012) Energetic and exergetic analyses of a dual-fuel diesel engine. Renew Sustain Energy Rev 16(7):4651–4660
13. Lopez I, Quintana CE, Ruiz JJ, Cruz-Perag F, Dorado MP (2014) Effect of the use of oliveep-omace oil biodiesel/diesel fuel blends in a compression ignition engine: preliminary exergy analysis. Energy Convers Manage 85:227–233
14. Aghbashlo M, Tabatabaei M, Mohammadi P, Pourvosoughi N, Nikbakht AM, Goli SAH (2015) Improving exergetic and sustainability parameters of a DI diesel engine using polymer waste dissolved in biodiesel as a novel diesel additive. Energy Converse Manage 10(5):328–337
15. Caliskan H, Tat ME, Hepbasli A, Gerpen JHV (2010) Exergy analysis of engines fuelled with biodiesel from high oleic soybeans based on experimental values. Int J Exergy 1(7):20–36
16. Sayin C, Hosoz M, Canakei M, Kilicaslan I (2007) Energy and exergy analysis of a gasoline engine. Int J Energy Res 31(3):259–273
17. Sekmen P, Yilbasi Z (2011) Application of energy and exergy analyses to a CI engine using biodiesel fuel. Math Comput Appl 6(4):797–808
18. Biplab KD, Sahoo N, Ujjwal KS (2013) Thermodynamic analysis of a variable compression ratio diesel engine running with palm oil methyl ester. Energy Conserv Manag 65:147–154
19. Panigrahi N, Mohanty MK, Acharya SK, Mishra SR, Mohanty RC (2014) Experimental investigation of karanja oil as a fuel for diesel engine—using shell and tube heat exchanger, World Academy of Science, Engineering and Technology. Int J Chem Mat Sci Eng 8(1):91–98
20. Panigrahi N, Mohanty MK, Mohanty RC, Mishra SR (2016) Performance of a C.I. engine with energy and exergy analysis fuelled with neem oil methyl ester. Int J Renew Energy Technol 7(3):264–287
21. Moran MJ, Shapiro HN, Boettner DD, Bailey M (2014) Fundamentals of engineering thermodynamics, eighth edn. Wiley
22. Cengel YA, Boles MA (2009) Thermodynamics: an engineering approach, 6th edn. Tata Mc GRAW Hill Companies, New Delhi
23. Canakci M, Hosoz M (2006) Energy and exergy analysis of a diesel engine fuelled with various biodiesels. Energy Sour Part B 1:379–394
24. Shamshirband S, Tabatabaei M, Aghbashlo M, Yee L, Petkovi D (2016), Support vector machine-based exergetic modelling of a DI diesel engine running on biodiesel blends containing expanded polystyrene. Appl Therm Eng 727–747
25. Lopez I, Quintana CE, Ruiz JJ, Cruz-Perag F, Dorado MP (2014) Effect of the use of oliveep-omace oil biodiesel/diesel fuel blends in a compression ignition engine. Preliminary exergy analysis. Energy Convers Manage 227–233
26. Rakopoulos CD, Giakoumis EG (2006) Second-law analyses applied to internal combustion engines operation. Prog Energy Combust Sci 31(12):2–47

Mr. Amar Kumar Das is pursuing as PhD Scholar in the Department of Mechanical Engineering, VSSUT, Burla. He has been both graduated in Mechanical Engineering in the year 2003 and completed his MTech in Heat power engineering in 2011 from BPUT, Rourkela. He has a good record of academic excellence and teaching experience of around 12 years. He has been involved in many projects in waste management and Renewable energy and being awarded by government and social organisations. He is also continuing as life member in many professional societies of national repute like ISTE, IEI, OBA and SMAE. He has published 10 papers on waste management and renewable energy. He has been involved in conducting different seminar, FDP and conferences.

Dr. Achyut K. Panda is a faculty member in the Department of Chemistry at the Veer Surendra Sai University of Technology Burla, Odisha, India. He completed his Ph.D. from the Department of Chemical Engg. National Institute of Technology Rourkela and has more than 18 years of teaching experience. His research interest includes waste plastics to fuel, bio energy and catalysis. He has more than 40 research publications in different reputed journals and conferences and one Indian patent in his credit. Most of the published articles are related to the value added utilisation of different types of waste plastics and non-edible biomass through thermochemical technique. In addition to this, he also has published book and popular articles, organised conferences and faculty development programme and is a reviewer of scientific manuscripts from different journals.

Dr. (Mrs.) Dulari Hansdah is working as an Assistant Professor in the Department of Mechanical Engineering, at NIT Jamshedpur. She graduated in Mechanical Engineering from Indira Gandhi Institute of Technology (IGIT), Sarang, Dhenkanal in the year 2008. She did her postgraduate study in Mechanical Engineering with specialization in Thermal Engineering at National Institute of Technology (NIT) Rourkela. Immediately after completion of M.Tech in 2011, she obtained her Ph.D. from NIT Rourkela in the year 2016 and her areas of research are alternative fuels for IC engine, renewable energy, engine emissions and control methods, engine technology for carbon neutral and waste utilisation from various Sectors. She has developed the electronically controlled unit (ECU) for HCCI mode of operation of an engine during her PhD and designed the vaporiser for bioethanol fumigation. She utilised the ECU system and vaporiser for experiment available in NIT Rourkela. She has 16 publications (two SCI journal papers, four Scopus indexed journals and ten conference proceedings) in the field of experimental, computational modelling and analysis of engine combustion and emissions. Apart from this, she is also well verged with different modelling and FE software such as ANSYS, SolidWorks and openFOAM. She is the reviewer of progress in industrial ecology (PIE), an international journal of inderscience publishers and also the reviewer of international journal of automotive and mechanical engineering.

Thermal Analysis and Performance Evaluation of Peltier Module

Sumit Kumar and Dhaneshwar Mahto

Abstract In day-to-day life, the air conditioning industry is growing due to climate change and it affects adversely to the environment. Power consumption is a vital problem in air conditioning industry. Air condition industry affects the environment by using different refrigerant. It affects the ozone layer. Refrigerants effects on the environment are unpredictable. So, scientists and researchers are looking for an environmental friendly air conditioning system. In the concern of environment cooling using Peltier effect are viable. Peltier element directly produced a cooling effect by using electrical energy. This system does not use any refrigerant so it is the eco-friendly system. Global warming increases rapidly with the social development. To reduce these impacts, thermoelectric cooling systems are suggested as one of the promising technology in air conditioning industry. Also, the rising costs of energy and increasing environmental awareness have given enough focus on the use of the thermoelectric module. Advantages of thermoelectric cooling systems are, compact in size, required less space, less weight, high reliability, the absence of moving parts and no working fluids are required. It also possesses the advantage that it can be run by Direct Current (DC) which can be run by solar energy (Photovoltaic cells). In the present scenario, thermoelectric technology is utilizing in the spaces which required a precise temperature control. Despite many advantages, the Peltier element has only a very small market share in the field of refrigeration because of low efficiency and high material cost. In the present work thermal and performance evaluation of 191 W Peltier modules has been carried out. It is found that at certain current strength the cooling capacity stops increasing and decreased instead. Also, heating output increases parabolically and never reaches zero for the positive value of '*I*'.

Keywords Peltier module · Seebeck effect · Global warming
Thermoelectric generator

S. Kumar (✉) · D. Mahto
Department of Mechanical Engineering, Birla Institute of Technology, Mesra,
Ranchi 835215, Jharkhand, India
e-mail: sumitsahu0204@gmail.com

J. Chattopadhyay et al. (eds.), *Renewable Energy and its Innovative Technologies*,
https://doi.org/10.1007/978-981-13-2116-0_14

Nomenclature

COP	Coefficient of performance
CFC	Chlorofluorocarbon
HFC	Hydro fluorocarbon
$\dot{Q}_h$	Heating capacity in Watt
$\dot{Q}_c$	Cooling capacity in Watt
$\dot{Q}_j$	Joule heat
I	Electric current in Ampere
V	Voltage in volts
R	Resistance in ohm
P_{el}	Power of Peltier element
T_h, T_c	Heating and cooling temperature in °C

1 Introduction

Energy crisis and global warming have become more and more serious with the social development. Due to greenhouse gas emission, ozone layer gets depleted and also terminologies like acid rain, global warming appears.

The surface temperature of Earth has risen about 1.1 °C from last 25–30 years; this change is due to increased carbon dioxide emissions into the atmosphere [1]. Global warming has increased most in the past 16 years. 2016 was the warmest year; most of the months of this year were warmest comparative to other years [2].

Previously CFCs were used as a refrigerant for cooling. As these are volatile organic compounds, they may react with the gases which are present at the ground level to produce ozone, it can damage plants and materials health on a local scale. At a global scale, the release of CFCs to the environment has serious environmental effects. The lifespan of CFC is long in the atmosphere which leads it to get involved in the reaction with the ozone layer which is present in Stratosphere and converts the ozone into oxygen molecule [3]. Thus depleting the ozone layer and due to this, a hole is created which is a direct passage to ultraviolet rays. These radiations are harmful to the organism as well as environment. Now HFCs are used as a refrigerant. At present we cannot say whether HFCs are safe or not. It may be possible that after 10–20 years very serious environmental issues will be detected due to use of these refrigerants. So it's better to go for alternatives for the cooling system so that we can reduce the global warming problem.

To reduce these impacts, thermoelectric energy conversion device is proposed as one of the preferred technology in air conditioning industry.

Also, the rising costs of energy and increasing environmental awareness have given enough focus on the use of the thermoelectric module. This module works on Peltier effect which enables direct conversion between thermal and electrical energy. Peltier elements are electro-thermal converters that can be used as heat pumps

[4]. Physically they operate according to the Peltier effect, which corresponds to the inverse of the well-known Seebeck effect. This effect describes an electrical current in an element causing heat transfer from the cold to the warm side. The direction of the heat transfer depends on the current direction. Therefore, reversing the polarity changes the direction of heat transfer. In order to achieve the highest possible efficiency, the materials used in these elements have high electrical conductivity and consequently very low thermal conductivity [5]. This is why semiconductors such as bismuth telluride, antimony telluride or bismuth selenide are used to make thermoelectric module [6].

2 Literature Review

Bansal et al. [7] compared the performance of all three, i.e., vapor compression system, vapor absorption system, and thermoelectric system of similar capacity. It is found that vapor compression consumes least energy was least costly but was the noisiest. Absorption refrigerator was quietest of the three but most expensive. Thermoelectric was costliest among all.

Min et al. [8] experimentally investigated the prototype thermoelectric refrigerator. COP of thermoelectric refrigerators is found to be around 0.3–0.5 at an operating temperature of 5 °C. The cooling capacity of the thermoelectric module is also investigated. And the result showed COP can be improved by upgrading the module contact resistance, improving the thermal interfaces and by increasing the effectiveness of heat exchangers.

Manohar et al. [9] compared the performances of vapor compression system and the thermoelectric module. The result showed, in the freezer compartment temperature of the water decreased linearly with increase in time for compression cycle. And for thermoelectric refrigerator, the temperature of water decreased exponentially with increase in time. It means that cooling rate decreased exponentially for the thermoelectric system. For the majority of the cooling time, the cooling rate of the thermoelectric refrigerator was faster than the commercial refrigerator.

Jeong [10] theoretically investigated the process to optimize the thermoelectric module.

To maximize the value of COP for a thermoelectric module, the optimum current is determined with the help of cooling capacity of a thermoelement, the temperatures of the hot and cold side, both thermal and electrical resistance and the characteristics of the thermoelectric material.

Zhao et al. [5] analyzed the recent advances of the thermoelectric system and also reviewed the thermoelectric cooling. They also described techniques to improve thermoelectric cooling system's performance and inspected thermoelectric cooling applications for a different type of refrigeration including domestic refrigeration, electronic cooling, scientific and automobile applications.

He et al. [1] discussed the basic concept of thermoelectric refrigeration systems and its recent material. They concluded that this technology can be used in many areas with the advancement in the figure of merit and the device properties. It can be implemented in domestic air conditioning system and also used for power generation.

Ibanez-Puy et al. [11] developed different tests under real conditions to estimate the performance of the new thermoelectric system and also COP of this module. They Studied thermoelectric module for both cooling and heating mode and found that it has great advantages over commercial refrigeration for use in buildings as it can provide cooling and heating at the same time with the same machine.

3 Working Principle

One semiconductor is p-doped and the other n-doped, with the result that both have different conduction bands. When the voltage source is applied to the thermoelectric module, electrons start to flow. Electrons can move freely in conductors but cannot flow freely in the semiconductor as compared to conductors. When electrons enter the p-type since holes are present in p-type electrons fill a hole. Electrons drop from the upper conduction band to lower conduction band and release energy in the process. Then, electrons move from the p-type to the metal plate, the electrons are jumped up to upper conduction level and absorb energy in the process. Next, when the electrons flow into the n-type semiconductor, electrons jumped up to the upper conduction band and energy is absorbed in this process. At last, when the electrons leave the n-type semiconductor and enter the metal plate, they drop down from the upper conduction band to lower conduction band and release energy in the process [3]. It means that side on which heat is absorbed is cold side and side on which heat is released is hot side (Fig. 1).

Since the conduction band of n-doped semiconductor material has a lower energy level, the cooling takes place at the point where the electrons pass from the n-doped conductor to the p-doped one. In a practical thermoelectric module, two or more

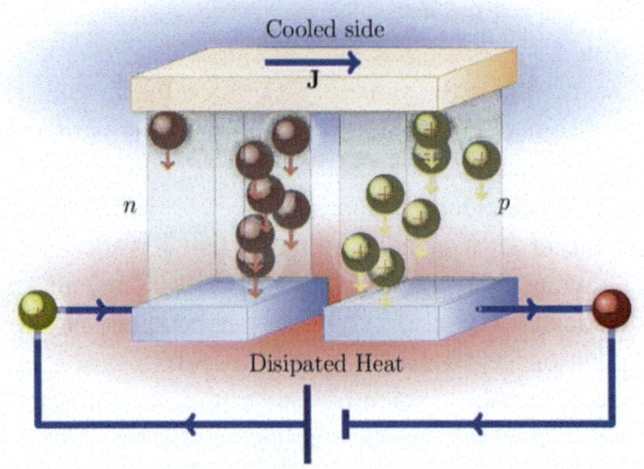

Fig. 1 Thermoelectric cooling process

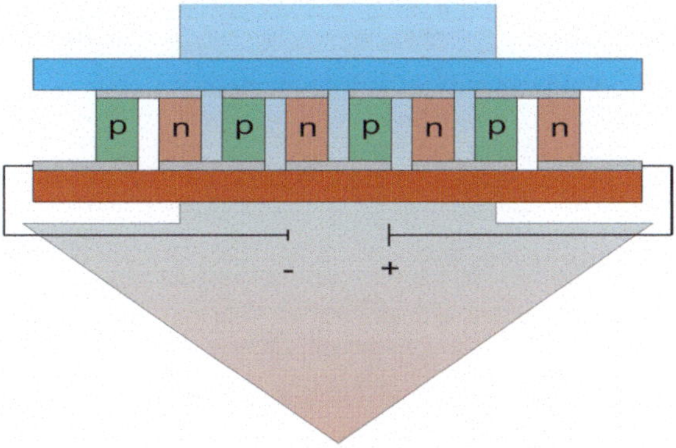

Fig. 2 Thermal current in Peltier element

PN semiconductor couples are used which are connected electrically in series but thermally in parallel (Fig. 2).

3.1 *Thermoelectric Behavior*

According to the first law of thermodynamics, energy cannot be destroyed, only converted into another form of energy. There are two energy flows acting in the Peltier element; the cooling capacity and the electrical output. These two energy flows must be dissipated into the environment via the heat lost from the hot side. Thus, the energy balance of the element can be expressed as [11]:

$$\dot{Q}_h = P_{el} + \dot{Q}_c \tag{1}$$

The quantity of heat transported by the Peltier effect behaves proportionally to the flowing current 'I'. Furthermore, it is dependent on the temperature of the cold side T_c and the Seebeck coefficient α [12].

$$\dot{Q}_p = \alpha.T_c.I \tag{2}$$

Two types of loss occurred on Peltier element.

- Joule heat due to ohmic resistance

$$P_{el} = \dot{Q}_j = I^2R \tag{3}$$

- A part of the cooling capacity is lost through the energy flow from the warm to the cold side. This depends on the thermal conductivity (Λ) and the temperature difference (ΔT) of the element.

$$\dot{Q}_{\Delta T} = \Lambda.\Delta T \tag{4}$$

Cooling capacity of Peltier element is

$$\dot{Q}_{\mathrm{c}} = \alpha.I.T_{\mathrm{c}} - \frac{1}{2}R.I^2 - \Lambda(T_{\mathrm{h}} - T_{\mathrm{c}}) \tag{5}$$

Heating capacity is

$$\dot{Q}_{\mathrm{h}} = \alpha.I.T_{\mathrm{h}} - \frac{1}{2}R.I^2 - \Lambda(T_{\mathrm{h}} - T_{\mathrm{c}}) + P_{\mathrm{el}}$$

Therefore,

$$\dot{Q}_{\mathrm{h}} = \alpha.I.T_{\mathrm{h}} + \frac{1}{2}R.I^2 - \Lambda(T_{\mathrm{h}} - T_{\mathrm{c}}), \tag{6}$$

where $\alpha = \frac{V_{\mathrm{max}}}{T_{\mathrm{h0}}}$

$$R = \frac{(T_{\mathrm{h}} - \Delta T_{\mathrm{max}}).V_{\mathrm{max}}}{T_{\mathrm{h0}}.I_{\mathrm{max}}}$$

$$\Lambda = \frac{(T_{\mathrm{h}} - \Delta T_{\mathrm{max}}).V_{\mathrm{max}}.I_{\mathrm{max}}}{2T_{\mathrm{h0}}.\Delta T_{\mathrm{max}}}$$

COP indicates the ratio of net power to the power fed in. Net power is the cooling capacity.

$$\mathrm{COP} = \frac{\dot{Q}_{\mathrm{c}}}{P_{\mathrm{el}}} \tag{7}$$

3.2 THCU (Thermoelectric Cooling Heating Unit)

Numbers of PN semiconductor couples are connected electrically in series and semiconductors are connected through metal bridge usually copper. These are arranged in small blocks between two ceramic plates. Materials with the lowest possible electrical resistance at high temperature are used (Fig. 3).

4 Process Description

Applying an electric voltage causes one side of the Peltier element to become warm and the other side to become cold. The resulting thermal currents are dissipated in

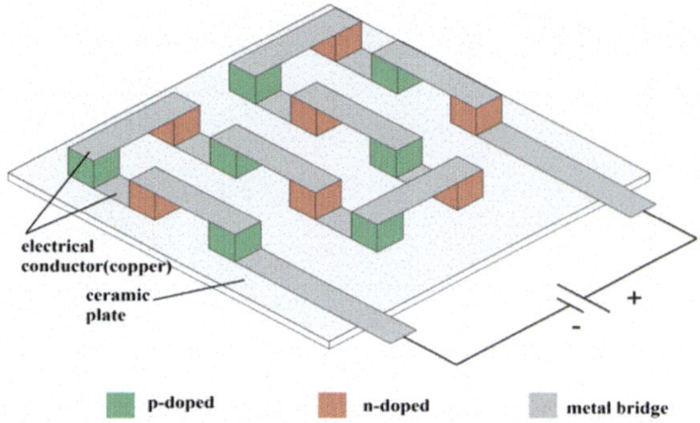

Fig. 3 Construction of Peltier module

the system by water-cooled heat exchangers. Water-cooled heat exchangers have an advantage compared to air-cooled ones, in that firstly the heat transfer is considerably cheaper and second, the energy flows can be easily measured. To determine the energy flows, the water mass flow rate must be multiplied by the temperature change across the heat exchanger and the specific thermal capacity. In order to investigate the behavior of a Peltier element, a potentiometer can be used to adjust the voltage drop across the element.

Additionally, the water mass flow can be changed by using the adjustment screws on the flow meters. This makes it possible to influence the surface temperature on the elements, which also has an effect on performance. Since the heating output of the Peltier element is greater than its cooling capacity, running the experiment for a longer period would heat up the water in the tank. To prevent this, there is another heat exchanger located in the supply tank, which is used to cool the water flowing back to the tank. Cooling here is done by the laboratory water supply.

4.1 Process Diagram

See Fig. 4.

5 Results and Discussion

T_1 Temperature of cold side,
T_2 Temperature of hot side,
T_3, T_4 Temperatures of inlet and outlet to cold side heat exchanger,
T_5, T_6 Temperatures of inlet and outlet to hot side heat exchanger (Table 1).

$$\text{Cooling capacity} = (\rho C_\mathrm{p})_\mathrm{water} * \text{Volume flow rate of water} * (T_3 - T_4)$$

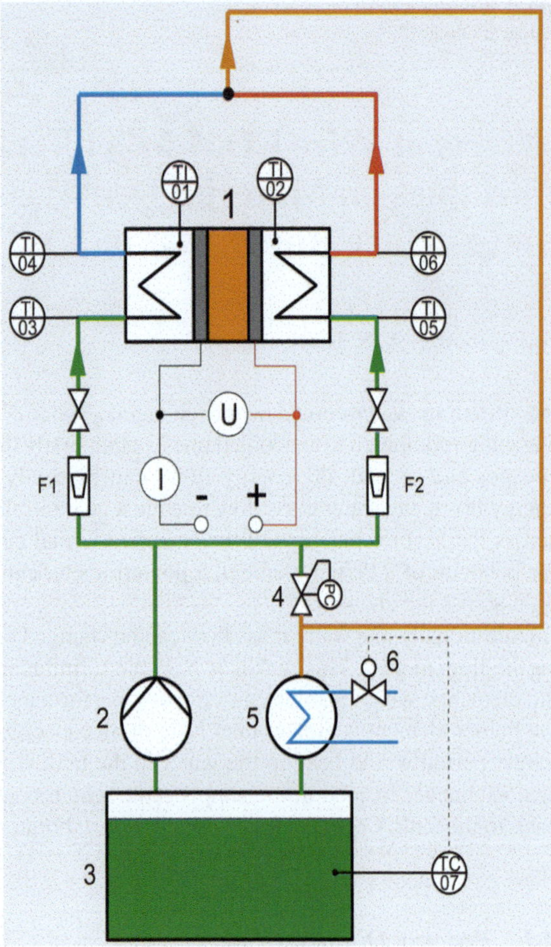

Fig. 4 Process diagram of Peltier module with heat exchangers

$$\text{Thermal output} = (\rho C_{\text{p}})_{\text{water}} * \text{Volume flow rate of water} * (T_6 - T_5)$$

Figure 5 shows the effect of voltage on cooling capacity. Cooling capacity first increases with the increase of voltage. After reaching the maximum value cooling capacity decreases due to increase of Joule heat and larger temperature difference in the hot and cold side of Peltier elements.

The voltage at which cooling capacity reaches maximum value is threshold voltage. If the voltage exceeds from threshold limit, the cooling capacity drops rapidly. This is due to following facts:

1. The resulting Joule heat increases with the square of the applied voltage. Above the certain voltage, this effect has a prominent in thermal current than the Peltier effect.

Table 1 Temperature and current readings at different voltage

V (V)	I (A)	T_1 (°C)	T_2 (°C)	T_3 (°C)	T_4 (°C)	T_5 (°C)	T_6 (°C)
2	1.9	21.4	25.9	23.4	21.7	22.8	23.6
3	2.9	20.8	27.3	23.5	21.4	22.4	23.6
4	3.9	19.9	29.6	23.8	20.6	23.6	24.4
5	4.8	19.1	31.2	23.9	19.8	22.9	24.8
6	5.8	18.4	32.7	23.9	19.4	23.5	24.9
7	6.7	17.9	34.5	23.8	19	23.6	25.3
8	7.6	17.4	36.2	23.8	18.8	23.6	25.6
9	8.4	17.1	38.2	23.6	18.5	23.6	26
10	9.2	17.2	38.9	23.4	18.6	21.8	25.1
11	10	16.8	42	23.2	18.2	23.7	26.8
12	10.8	17.1	43.8	23.3	18.6	23.4	26.8
13	11.7	16.8	45.3	23.3	18.3	23.8	27.5
14	12.4	17	47.8	23.7	18.3	23.6	28.1
15	13.1	16.9	50	23.4	18.5	23.8	28.3
16	13.9	17	51.9	23.5	18.6	23.9	28.8
17	14.4	17.4	53.8	23.5	18.7	23.7	29.5
18	15.2	17.9	55	23.8	19.3	22.1	28.8
19	15.8	18.2	56.7	23.9	19.5	22.1	29.4
20	16.5	18.5	58.7	23.8	19.6	22.2	30
21	16.9	19	60.4	23.8	19.9	22.6	30.2
22	17.5	19.7	62.3	23.7	20.6	22.4	30.9
23	18.1	20.5	63.9	23.5	21	23.3	31.5
24	18.4	20.8	64.8	23.6	21.2	22.5	31.9

Fig. 5 Variation of cooling capacity with applied voltage

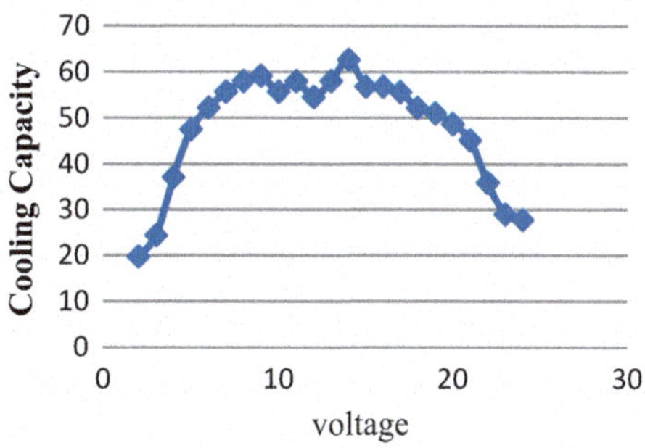

2. The temperature difference between warm and cold side also rises with increasing voltage, which also causes the higher losses.

Figure 6 shows the effect of voltage on the thermal output. Thermal output increases with increase in voltage. This is due to the reason of increase of Joule heat and increase in temperature of the heating plate.

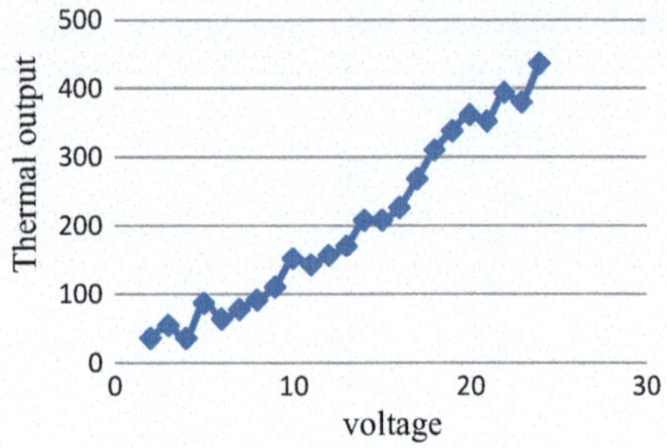

Fig. 6 Variation of thermal output with applied voltage

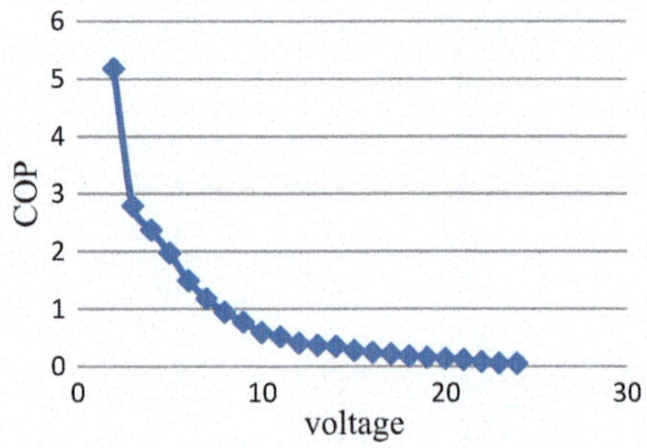

Fig. 7 Variation of COP with applied voltage

Figure 7 shows the variation of the coefficient of performance with voltage. With the increase in voltage, COP decreases asymptotically. Theoretically, it approaches a value of −0.5 for an infinite voltage. Therefore from an energy point of view, it is cheaper to operate Peltier elements at low voltage. However, it should be noted that the cooling capacity decreases sharply below the ideal voltage. Since efficiency only plays a secondary role in most usual applications, the elements are normally operated around the ideal voltage value. From Fig. 7 coefficient of performance is 0.32 at the threshold voltage of 13 V which is found from experimental setup.

6 Conclusion

In the present work Peltier element module of 191 W capacities is experimentally investigated and found the followings:

- Cooling capacity decreases after reaching threshold voltage.
- COP is 0.32 at the threshold voltage of 13 V.
- Peltier element gives high performance at low voltage.

References

1. He W, Zhang G, Zhang X, Ji J, Li G, Zhao X (2015) Recent development and application of thermoelectric generator and cooler. Appl Energy 143:1–25
2. Potter S, Cabbage M, McCarthy L (2017) NASA, NOAA data show 2016 warmest year on record globally. National Aeronautics and Space Administration (NASA)
3. Rawat MK, Chattopadhyay H, Neogi S (2013) A review on developments of thermoelectric refrigeration and air conditioning systems: a novel potential green refrigeration and air conditioning. Int J Emerg Technol Adv Eng 3(3):362–367
4. Hamid Elsheikh M et al (2014) A review on thermoelectric renewable energy: principle parameters that affect their performance. Renew Sustain Energy Rev 30:337–355
5. Zhao D, Tan G (2014) A review of thermoelectric cooling: materials, modeling and applications. Appl Therm Eng 66(1–2):15–24
6. Gaikwad M, Shevade D, Kadam A, Shubham B (2016) Review on thermoelectric refrigeration: materials and technology. Int J Curr Eng Technol (INPRESSCO IJCET) 4(4):67–71
7. Bansal P, Martin A (2000) Comparative study of vapour compression, thermoelectric and absorption refrigerators. Fuel Energy Abstr 41(5):320
8. Min G, Rowe DM (2006) Experimental evaluation of prototype thermoelectric domestic-refrigerators. Appl Energy 83(2):133–152
9. Manohar K, Adeyanju AA (2014) Comparison of the experimental performance of a thermoelectric refrigerator with a vapour compression refrigerator. Int J Tech Res Appl 2(3):1–5
10. Jeong ES (2014) A new approach to optimize thermoelectric cooling modules. Cryogenics (Guildf) 59:38–43
11. Ibañez-Puy M, Bermejo-Busto J, Martín-Gómez C, Vidaurre-Arbizu M, Sacristán-Fernández JA (2017) Thermoelectric cooling heating unit performance under real conditions. Appl Energy 200:303–314
12. Cosnier M, Fraisse G, Luo L (2008) An experimental and numerical study of a thermoelectric air-cooling and air-heating system. Int J Refrig 31(6):1051–1062

Sumit Kumar is a M.E. student at department of Mechanical Engineering, **Birla Institute of Technology Mesra**, **Ranchi**, **India**. He has completed his B.Tech. from CEB, Bhubaneswar.

Dr. Dhaneshwar Mahto is an Assistant Professor at Mechanical Engineering department, **Birla Institute of Technology Mesra, Ranchi, India**. He has completed his Ph.D. from SHIATS, Allahabad. His research interest includes Refrigeration and Air conditioning and Non-conventional energy.

Experimental Investigation of Pumpless Vapour Absorption System

Vikash Kumar, Sumit Kumar and Dhaneshwar Mahto

Abstract Energy is the driving force of civilization. It is also a measure of the prosperity of the nation. In the present day, per capita energy consumptions increase with the increase of income. Developed countries consume more energy compared to developing countries. About 10–20% of energy has been consumed in air conditioning industry. Vapour absorption system (VAS) reduces power consumption and causes no ozone depletion. It consumes low grade of energy with respect to vapour compression system, which consumes high grade of energy. The present work experimentally analyzed the 25 bar and 230 V pumpless vapour absorption system.

Keywords Pumpless vapour absorption system · Water-ammonia-hydrogen Generator · Carrier gas

1 Introduction

In day-to-day life, the air conditioning industry is growing due to climate change and it also affects adversely to the environment. Power consumption is also a vital problem in air conditioning industry. The compressor is not used in pumpless vapour absorption system because the input of vapour absorption system is heat energy in place of mechanical energy. In place of compressor a generator, an absorber and a heat exchanger is used. In this system, ammonia is used as a refrigerant, water as absorbent and hydrogen as carrier gas. For the flow of solution from the absorber to generator pressure mechanical pump is used in conventional absorption refrigeration system. In pumpless vapour absorption system by using hydrogen as a carrier gas remove the pump. The pumpless vapour absorption system has several

V. Kumar · S. Kumar · D. Mahto (✉)
Department of Mechanical Engineering, Birla Institute of Technology, Mesra, Ranchi 835215, Jharkhand, India
e-mail: dhaneshbit@yahoo.co.in

V. Kumar
e-mail: contacttovikash@gmail.com

J. Chattopadhyay et al. (eds.), *Renewable Energy and its Innovative Technologies*,
https://doi.org/10.1007/978-981-13-2116-0_15

advantages over conventional system vapour absorption system like high reliability due to absence of moving parts, low maintenance and silent operation [1].

Vapour absorption system is mostly used in food storage in cold store and also in food transporting vehicle. It may be used a waste heat from vehicle to run the vapour absorption system. It can also use in industry was the maximum waste heat is produced like in gas turbine, and steel industry, etc. The vapour absorption refrigeration system can be used in urban as well as rural areas. Its main advantage is that it may utilize low grade of energy in the generator. So, it may viable in remote areas where no electricity is available.

2 Literature Review

Chen et al. [2] proposed a new generator in vapour absorption refrigeration to increase the COP up to 50%.

Saravanan et al. [3] analyzed heat operated bubble pump vapour absorption refrigerator with air cooled condenser and absorber. Their examination uncovers that the impact of the evaporator to absorber pressure drop on the framework execution is more noteworthy than that of the generator to condenser pressure drop.

Koyman et al. [4] performed tentatively R22 as option working medium. They found energetic efficiency is 0.35.

Ma et al. [5] investigated two fluid models of two-stage flow and heat transfer process in a two-stage fastened thermo syphon. They revealed that flow designs predicated numerically and dissemination of parameters under various working conditions demonstrated a decent concurrence with test comes.

Zohar et al. [6] have been study the feasibility of R32, R124, R125 and R134A as refrigerants in vapour absorption system. They found NH_3/H_2O COP is higher than there mixture.

Mazoir et al. [7] investigated tentatively a commercial diffusion absorption refrigeration system machine to determine its performance. They found that best performance of the machine was with a heat supply of 42 W.

Rao et al. [8] have been analyzed Einstein–Slizard refrigeration cycle with solar energy as an input in a single pressure absorption cycle. They used driving mechanism in his study in two phase boiling phenomenon.

Enosoz et al. [9] experimentally investigated three distinctive heat inputs supplies to generator on the vitality execution of the diffusion absorption refrigeration system. To accomplish the objective, he considers three heat inputs 62, 80 and 115 W. He found that the highest energy performance is for DAR 62 W system as 0.36 and lowest for DAR 115 W system as 0.30.

Jemaa et al. [1] have been experimentally and theoretically analyzed vapour absorption system working with ammonia as refrigerant, water as absorbent and hydrogen as carrier gas. Their result indicates that the largest source of irreversibility is absorbed.

Jemaa et al. [10] performed an experiment analysis of diffusion absorption refrigeration machine. They figured worldwide heat transfer coefficient of the refrigerator and afterward discover cooling limit of the refrigerator. They found that the important electric energy of the warming component is in the range of 35–45 W.

3 System Description

3.1 *Function Diagram of the Pumpless Absorption Refrigeration System*

See Fig. 1.

3.2 *Absorption Refrigeration Process*

See Fig. 2.

3.3 *Working Principle*

The vapour absorption apparatus is made by a number of steel vessels and pipes are welded together to form a hermetically sealed equipment. The space of the apparatus is in open and unrestricted so that the pressure in all parts is same. The component of the system includes a generator a condenser, an evaporator, and an absorber. A diagram of the circuit is shown in Fig. 1.

Heat is applied by a gas burner or other source of heat to expel the ammonia from solution. The ammonia vapour generated and flow towards the condenser. In the flow of ammonia from the generator to condenser are interposed an analyzer and a rectifier. The analyzer and rectifier serve to remove this water vapour from the ammonia vapour. The ammonia vapour passes to the condenser where it is liquid by the air cooling. The ammonia thus liquefied flows into the evaporator. The liquid trap interposed between the condenser section and the evaporator to prevent hydrogen from entering the condenser. Hydrogen gas enters the lower evaporator section and flows upwards, in counter flow to the downward flow liquid ammonia. The lesser ammonia vapour pressure results in the evaporation of the ammonia, with consequent absorption and cooling of the surrounding which is in a well-insulated enclosure. The cool, heavy gas mixture of hydrogen and ammonia vapour formed in the evaporator leaves the top of the evaporator and passes downwards through the gas heat exchanger to the absorber. The difference in specific weight is sufficient to initiate and maintain circulation between the evaporator and the absorber. The gas heat exchanger transfer

Fig. 1 Functional diagram of the absorption refrigeration system

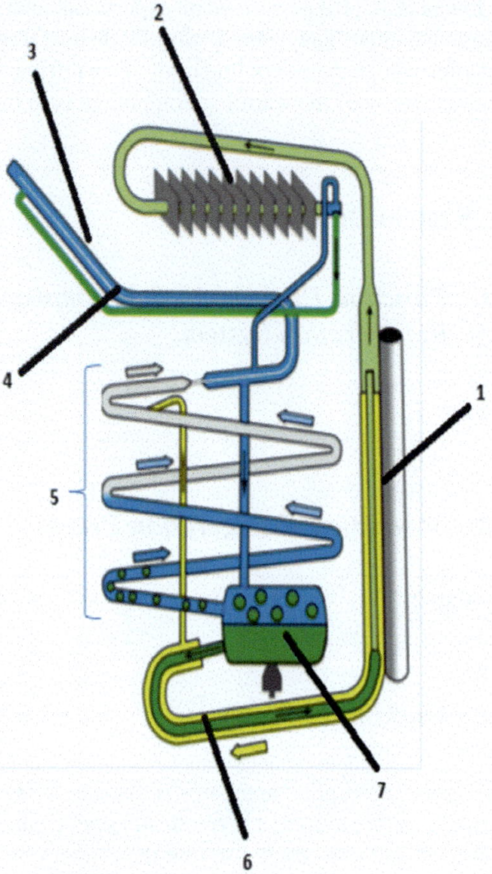

1 Boiler with vapour bubble
2 Condensers
3 Evaporator
4 Gas heat exchanger
5 Absorber
6 Liquid heat exchanger
7 Storage tank

Ammonia-rich solution

Weak ammonia solution

Ammonia/hydrogen gas mixture

heat from one gas to another. This saves some cooling in the evaporator by precooling the entering gas. In the absorber, a flow of weak solution comes in direct contact with the strong gas. Ammonia is absorbed by water, and the hydrogen passes towards from the top of the absorber through another chamber of the gas heat exchanger into the evaporator. The liquid and gas flow in opposite direction. The strong aqua ammonia solution flows through the liquid heat exchanger to the analyzer, and then to the strong liquid chamber of the generator. The liquid heat exchanger per-cools the liquid entering the absorber and preheat the liquid entering the generator. Refrigeration is accomplished by a continuation of the cycle.

3.4 Detailed Circuit Analyses

There are three distinct fluid circuits in the system an ammonia circuit, a hydrogen circuit includes the evaporator and absorber and a solution circuit including the generator and absorber shows in Fig. 2.

Fig. 2 Circuits of the absorption refrigeration process

Absorption refrigeration process

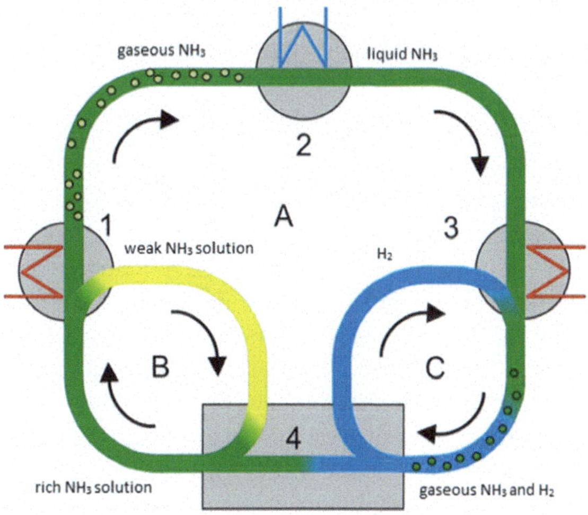

1 Boiler with vapor bubble pump

2 Condenser

3 Evaporator

4 Absorber

A Ammonia circuit

B Water circuit

C Hydrogen circuit

Water circuit: The storage tank contains a solution of ~65% water (H_2O) and ~35% ammonia (NH_3). This solution is known as the 'ammonia-rich solution'. It flows into the boiler from the storage tank. Ammonia vapour is vapourized from the solution at ~150–180 °C and passes up the riser pipe as bubbles. The vapour pressure pushes the solution through the liquid heat exchanger, in which the rich solution is preheated, and into the absorber. The weak solution drops through the pipe coils in the absorber and back into the storage tank.
Ammonia circuit: The ammonia vapour vapourized from the boiler reaches the condenser at ~70 °C. By the use of cooling fins, the condenser is kept at room temperature. In this temperature and a system pressure of ~25 bar, the ammonia vapour condenses. The liquid ammonia flows through the gas heat exchanger and into the evaporator, where it humidifies the inner surface of the evaporator. At that time, the hydrogen is blown in as a carrier gas.
Hydrogen circuit: The weak solution in the absorber absorbs the ammonia completely, while the hydrogen remains totally separated by these processes. The almost pure and lightened hydrogen gas exits the absorber at the upper end into the evaporator, where it is blown over the surface that humidified with ammonia. The temperature certainly drops to very low values as the liquid ammonia immediately evaporates into the hydrogen.

4 Result and Discussion

Figure 3 shows that the variation of temperature with time. The temperature of generator increase with the increase of time and after 25 min generator temperature is steady state, i.e. strong solution separated from an aqueous solution of ammonia. At the temperature of 140 °C of generator cooling of evaporator starts and reaches steady state at after 45 min. When evaporator reaches steady state the temperature difference between of evaporator and ambient temperature are maximum. This is due to the fact that after reaching steady state heat rejecter is maximum. Absorber and generator inlet temperature coincide with each other, because of weak solution and strong solution are mixed in the absorber and release the heat at ambient condition.

Figure 4 shows that variation of evaporator temperature with absorber temperature. Evaporator temperature decreases rapidly with the increase of absorber temperature and reaches minimum after that it increases with increase absorber temperature. This is due to the fact that with an increase in absorber temperature, heat rejected from strong solution and make solution mixture is decreased to the ambient temperature.

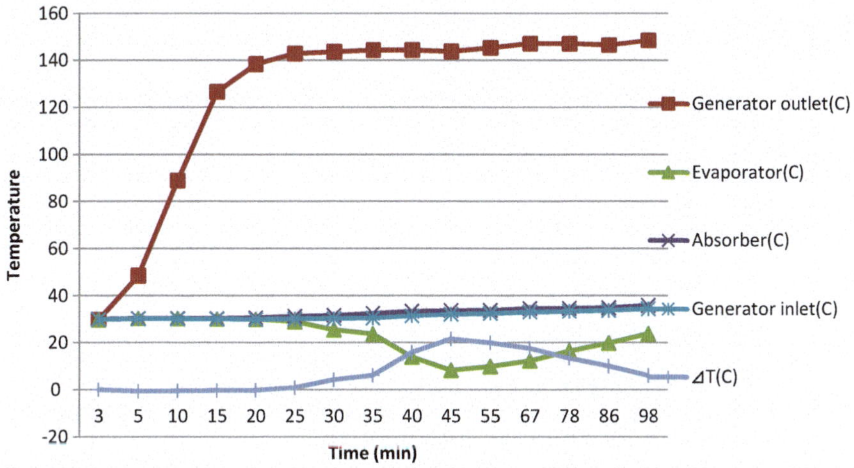

Fig. 3 Shows the variation of temperature with experimental time

Fig. 4 Variation of evaporator temperature with absorber temperature. *Note* (ΔT) Temperature difference between ambient temperature and evaporator temperature

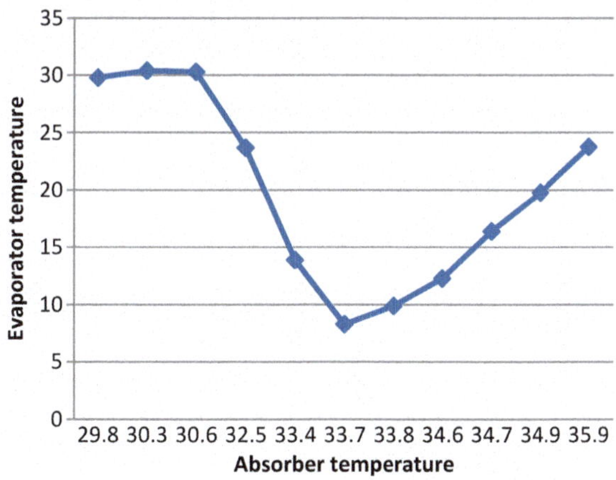

5 Conclusion

The following conclusions have been concluded by experimental results of pumpless vapour absorption system, using ammonia as a refrigerant, water as absorbent and hydrogen as carrier gas.

1. Pumpless vapour absorption system at (25 bar, 230 V) starts cooling at generator temperature 140 °C after 25 min.
2. Steady state reached the system after 45 min.
3. Steady state of vapour absorption system depends on ambient temperature.

References

1. Ben Jemaa R, Mansouri R, Boukholda I, Bellagi A (2016) Experimental investigation and exergy analysis of a triple fluid vapour absorption refrigerator. Int J Energy Convers Manag 124:84–91
2. Chen J, Kim KJ, Herold KE (1996) Performance enhancement of a diffusion absorption refrigerator. Int J Refrig 19:208–218
3. Saravanan R, Maiya MP (1997) Effect of component pressure drops in two fluid pump less continuous vapour absorption refrigerator. Energy Conserv 38:1823–1832
4. Koyfman A, Jelinek M, Levy A, Borde I (2003) An experimental investigation of bubble pumps performance for diffusion absorption refrigeration system with organic working fluid. Appl Therm Eng 32:1181–1194
5. Ma Z, Turan A, Guo S (2009) Practical numerical simulation of two phase flow and heat transfer phenomena in a thermosyphon for design and development. In: ICCS, Part 1, pp 665–674
6. Zohar A, Jelinek M, Levy A, Borde I (2009) Performance of diffusion absorption refrigeration cycle with organic working fluids. Int J Refrig 32:1241–1247
7. Mazouz S, Mansouri R, Bellagi A (2014) Experimental and thermodynamic investigation of an ammonia/water diffusion absorption machine. Int J Refrig 45:83–91
8. Rakesh Rao GN, Rahul G, Jayavardhan P (2015) Analysis of pump less vapour absorption refrigeration system. Int J Emerg Technol 9:69–72
9. Ersoz MA (2015) Investigation the effect of different heat inputs supplied to the generator on the energy performance in diffusion absorption refrigeration system. Int J Refrig 54:10–21
10. Ben Jemaa R, Mansouri R, Boukholda I, Bellagi A (2017) Experimental characterization and performance study of an ammonia-water-hydrogen refrigerator. Int J Hydrogen Eng 42:8594–8601

Vikash Kumar is a ME (Heat Power) students of **Birla Institute of Technology Mesra, Ranchi, Jharkhand, India**.

Sumit Kumar is a M.E. student at department of Mechanical Engineering, **Birla Institute of Technology Mesra**, **Ranchi**, **India**. He has completed his B.Tech. from CEB, Bhubaneswar.

Dr. Dhaneshwar Mahto is an Assistant Professor at Mechanical Engineering department, **Birla Institute of Technology Mesra**, **Ranchi**, **India**. He has completed his Ph.D. from SHIATS, Allahabad. His research interest includes Refrigeration and Air conditioning and Non-conventional energy.

Recent Trends in Application of Electrical Resistivity Tomography for Landslide Study

Philips Omowumi Falae, D. P. Kanungo, P. K. S. Chauhan and Rajesh Kumar Dash

Abstract Landslide disasters are known to be complex geological event with high socioeconomic effects in terms of loss of lives and properties. Landslide study involves the application of a multidisciplinary approach. Various techniques involving both surface and subsurface methods have been applied to investigate and understand the mechanism of its occurrence. Application of geophysical tools is one of such advanced techniques. In this paper, the recent trend in application of Electrical Resistivity Tomography (ERT) technique is being studied as it is a noninvasive method for investigation. ERT technique has been used widely for various landslides from rockfall to debris flow. The method has been applied to various geological conditions and environment to identify the slip surface, hydrological conditions, reconstruction of the slope body, estimation of the sliding materials volume/thickness, and depict the internal structures. Various array used in the measurement along with the algorithms for topographic inversion are also discussed. The results of 2D, 3D and 4D ERT techniques used in the recent times have been discussed. Future prospects and the limitations are also discussed.

Keywords Landslide · Electrical resistivity · Tomography · Array

1 Introduction

The earth surface in hilly region is constantly changing due to neo-tectonic activity and infrastructural developmental activities. Landslide occurs when the topsoil loses its strength by increase in moisture content or gravitational force or pull. It can be in the form of a rockfall or massive mudflow, but wherever the ground fails, it can

P. O. Falae (✉) · D. P. Kanungo · P. K. S. Chauhan · R. K. Dash
Geotechnical Engineering Group, CSIR-Central Building Research Institute, Roorkee 247667, Uttarakhand, India
e-mail: philomo08@gmail.com

P. O. Falae · R. K. Dash
Academy of Scientific and Innovative Research (AcSIR), Ghaziabad, India

J. Chattopadhyay et al. (eds.), *Renewable Energy and its Innovative Technologies*,
https://doi.org/10.1007/978-981-13-2116-0_16

result in massive destruction. Landslide disasters are known to be complex geological event with high socioeconomic effects in terms of loss of lives and properties [1]. The study of landslide requires integration of various disciplines and techniques in order to derive significant information, therefore, overcoming the drawback and limitations of applying single approach. The application of geophysical tools in the study of slopes and landslides has widely increased. Therefore, geophysical methods like seismic, electrical, and electromagnetic have got great attention in the recent times [1, 2].

Recently, the technique of ERT has metamorphoses from the primate methods of data collection, coverage as well as data processing. This improvement from the primitive sounding techniques has helped in overcoming poor resolution of complex geological terrains [3, 4]. The method has widely been used as a tool for investigating slope failures and landslides [1]. ERT technique has been used with an aim to reconstruct the slope body geometry, to identify the possible sliding surface, to identify the weak zone, estimation of the thickness of the sliding materials, and location of areas with high moisture content [1, 5, 6]. ERT also provides the information regarding the physical properties, materials of the landslide [1, 7, 8]. Newer algorithms for tomography data inversion, robust models for describing the hydro-geophysical processes have rapidly transformed the geoelectrical methods into a powerful and cost-effective tool for geo-hazard monitoring [9, 10].

The objective of this paper is to review the recent trends in application of ERT as a noninvasive method for landslide investigation and also to highlight the advantages as well as the limitations of the methods.

2 Literature Review

The ERT is an active geophysical method which can provide adequate information about electrical resistivity distribution of the subsurface. The results of electrical resistivity imagining have provided reasonable information in the identification of resistivity contrasts between the loss material and the bedrock which can be as a result of the nature of the Lithology and the degree of moisture content variation.

This method has widely been used by many researchers in the recent times to investigate various forms of landslide [1, 5, 6]. The properties of rock unit have been determined based on the electrical resistivity characterization of such rock unit. The rock mass has been found to depend largely on its physical properties such as porosity, pore fluid resistivity, and water saturation [8], as studied by Jomard et al. [9]. The true resistivity of a rock is also outlined using geophysical landslide investigation [10]. Landslide mass often have resistivity contrast different from the bedrock materials as a result of the losseness and deformation of the materials [9]. Therefore, a sharp resistivity contrast between the different masses might be an indication of slip surface, this should be easily identified. In the recent time, the application of ERT in landslide investigation has been improved from the conventional 1D to 4D imaging over time.

2D ERT imaging technique was used to investigate different typology of landslides ranging from shallow to deep-seated landslides [11, 12], debris flows [13,

14], rotational and transitional landslides [10, 15], slow moving slide, earth flow and quick clay [16, 17], complex landslides [18, 19], and rockslides [20, 21] while no case study of toppling failure has been reported.

The application of 3D imaging has been attributed to its capacity to determine landslide volumetric targets, reconstruction, and characterization. The invention of multichannels and multielectrode equipments has strongly increased the speed of data collection. It has been used by a number of researchers to study various types of landslides including complex landslides [22], translational and rotational slides [23], rock falls [24], and debris flows [25, 26].

The information provided by 2D and 3D imaging is static in nature due to their data acquisition techniques. The arrival of 4D imaging has given a room for continuous data collection. It has been applied in complex landslides [27], translational and rotational landslides [28, 29] and in debris flows [30].

3 Field Set Up Array

The field procedure comprises of the current multielectrode system, which is connected at a fixed distance according to the specification of electrode configuration on the ground. The current (*I*) is injected into the subsoil through the electrode and the voltage (*V*) is measured. The values of *I* and *V* depend on the geometric factor of the array configuration as indicated in Fig. 1 and the resistivity are calculated using Eq. (1) [31].

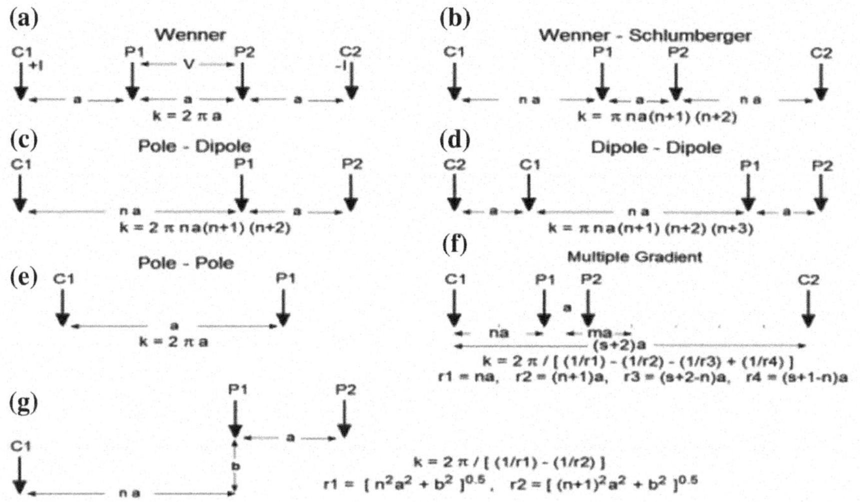

Fig. 1 Common ERT array and their geometric factors [3]

$$\rho = k\frac{\Delta V}{I} \quad (1)$$

where ρ is the resistivity, k is the geometric factor, I is the current flow, and Δv is the potential difference.

Several different arrays such as Schlumberger, Wenner, Wenner-Schlumberger, and Dipole-Dipole have been used by different researchers for carrying out the ERT study. Wenner-Schlumberger array achieves reasonable sensitivity to changes in both horizontal and vertical directions and has a higher signal-to-noise ratio than other configurations, as the potential electrodes are located inside the current electrodes [11, 32]. The Dipole-Dipole array is more prone to errors but highly sensitive to horizontal and vertical structures.

In some cases, multiple arrays are used along the same profile lines to measure the resistivity data for comparison analysis of the results and therefore, overcoming the drawback of any of the methods. It has been observed that Wenner-Schlumberger and Dipole-Dipole joint inversion method is the best combination for a better resolution and time of measurement [33], whereas Schlumberger and Dipole-Dipole combined inversion method gives a better resolution [3].

4 Recent Trends in ERT for Landslide Study

4.1 2D Electrical Resistivity Tomography

The application of 2D ERT techniques for landslide investigation is helpful in estimation of the overburden thickness, subsurface geological settings, geometry, depth of slip surface and tectonic set up. It also provides information about the groundwater conditions in terms of locating areas with high moisture content, the network of water drainages, and groundwater circulation regime as well as storage within an unstable area [3]. The usefulness of the ERT method to study the thickness of sliding layer, water accumulation zones, and the depth of the bedrock using the distribution of the true resistivity values on 2D profiles is also established [31].

4.2 3D Electrical Resistivity Tomography

The recent development on the representation of geological features in three-dimensional format has influenced the use of 3D resistivity surveying and interpretation method for landslide study. The application of 3D ERT method is helpful in carrying out the volumetric analysis, reconstruction, and characterization of the landslide body. The 3D methods are still not well explored because of their limitations in logistics such as transportation and installation of the instruments on uphill and downhill slopes [1]. This method is somehow costly in operation.

Recent developments of multielectrode and multichannel facility in such systems have enhanced their utility in data acquisition. Multichannel facility enables the user to acquire more than single reading/data at a particular time and is important in minimizing the survey time. Fast computing facility by providing in-built faster microcomputers in such systems enables the inversion of very large data sets [1].

4.3 4D Electrical Resistivity Tomography

Time-lapse ERT (4D ERT) surveys comprise of mobile, semipermanent or permanent electrodes with periodic measurements acquired over time and are increasingly popular in recent years [35]. These surveys have been applied across a wide range of environments and have recently been successfully applied to monitor active landslides in challenging transient natural systems [36]. One significant challenge in case of an active landslide is the possible movement of electrodes during monitoring and further studies have investigated techniques to compensate for potential errors [37]. In addition, a number of time-lapse inversion algorithms have recently been developed to provide improved time-lapse images [38]. Time-lapse ratio approach has been developed to give room to inversion models of the resistivity data acquired. Cascaded time-lapse inversion, difference inversion, and differencing in dependent inversions have also been developed [38]. 4D inversion with time dimension has also been developed [39].

5 Advantages and Limitations of ERT in Landslide Study

ERT has been widely used to delineate the internal structure of the landslide, for reconstruction of the slope body, to identify the slip surface, for estimation of the sliding materials (volume/thickness), and to delineate the area with high water content within the sliding body. Its application has also been proved to be very useful in situations whereby the resistivity contrast between the bedrock and the sliding body is interrelated to retrieve such information. This method of landslide investigation is cost-effective, noninvasive, and time effective.

The limitations in the application of ERT lie in areas with high contents of clayey materials or water. Presence of these materials can reduce the resistivity contrast, therefore making the interpretation hectic and difficulties in presenting the image of the topographic condition. To overcome these limitations, the correlation of ERT results with the results of other geophysical methods such as seismic methods, induced polarization (IP), ground penetration radar (GPR), electromagnetic method (EM), and spontaneous potential (SP) are the best option. Other parameters used in the correlation of ERT results can be geomorphology, borehole data, rainfall data, GPS

Table 1 Advantages and limitations of different ERT arrays used in landslide study

S. No.	Array	Advantages	Limitations	Reference
1	Wenner-Schlumberger array	• Sensitivity to changes in both horizontal and vertical directions • It has higher signal-to-noise ratio than other configurations • Good for shallow and noncomplex landslide	• It obtains narrower data than dipole-dipole array	[13, 22, 24]
2	Dipole-Dipole	• High sensitivity to horizontal changes in resistivity • Determination of isometric structures • Good for Deep-seated and Complex landslide	• Insensitive to vertical changes in the resistivity • More prone to errors because of very small signal strength	[39, 40]
3	Wenner	• Robust; high signal-to-noise ratio • Characterization of horizontal discontinuities • Good for shallow and noncomplex landslide	• Poor in the determination of vertical resolution • Susceptible to near-surface, lateral variations in resistivity which can be misinterpreted in terms of depth variations in resistivity	[41, 42]
4	Schlumberger	• Gives good mapping for the vertical change of the resistivity	• Very sensitive voltmeters are required • Limited to simple, horizontally layered structures	[15, 43]
5	Pole-Dipole Configuration	• To increase the depth of investigation it has relatively good horizontal coverage • Higher signal strength compared to dipole-dipole array • Lower sensitivity to telluric noise when compared with pole-pole array	• Lower signal strength when compared with Wenner and Wenner-Schlumberger arrays	[44]
6	Pole-Pole Configuration	• Widest horizontal coverage • Deepest depth of investigation	• Prone to large amount of telluric noise which can affect the quality of measurements • Poorest resolution	[45, 46]

data, inclinometer data, total station survey, stratigraphic information, and remote sensing data. This has helped in improving the interpretation of the ERT results [1, 3]. The advantages and limitations of ERT technique for landslide study are summarized in Table 1.

6 Conclusions

Application of ERT for landslide investigation has been explored and its advantages and the limitations are discussed. The application of ERT has provided the resistivity contrast of the boundary between the bedrock and the sliding body. This method is quite helpful in the identification of potentially stable and unstable slopes, high water content zone, geometry, and defining the geological settings in the landslide area. It is applicable for investigating all landslide typologies. In terms of array, Wenner array is good method in a noisy area to achieve a good vertical resolution with limited time of survey while dipole-dipole is recommended for good horizontal data coverage. Wenner-Schlumberger array is most suitable to obtain appreciable horizontal and vertical resolution with overlapping data levels, where there is limited number of electrode, pole-dipole array is appropriate taking measurements in both forward and reverse direction. Profiles with small electrode spacing and good horizontal coverage, pole-pole array are the best choice.

For better correlation and validation of ERT results, ground (surface and subsurface) information from various sources must be used.

Acknowledgements The authors wish to acknowledge the TWAS and CSIR for the doctoral fellowship granted to the first author to carry out his research work in CSIR-CBRI, Roorkee. We would like to thank the Director of CSIR-CBRI for granting permission for the review to be published.

References

1. Perrone A, Piscitelli S, Lapenna V (2014) Electrical resistivity tomography technique for landslide investigation: a review. Berichte Geol B-A 93:65–82. (ISSN 1017-8880—Applications in Landslide Monitoring)
2. de Bari C, Lapenna V, Perrone A, Puglisi C, Sdao F (2011) Digital photogrammetric analysis and electrical resistivity tomography for investigating the Picerno landslide (Basilicata region, southern Italy). Geomorphology 133:34–46
3. Loke MH, Chambers JE, Rucker DF, Kuras O, Wilkinson PB (2013) Recent developments in the direct-current geoelectrical imaging method. J Appl Geophys. https://doi.org/10.1016/j.jappgeo.2013.02.017
4. Mohd H, Abidin Z, Aziman M, Saiful A, Ahmad T, Mohd F (2017) Forensic assessment on near surface landslide using electrical resistivity imaging (ERI) at Kenyir Lake Area in Terengganu, Malaysia. Procedia Eng 171:434–444. Available online at www.sciencedirect.com
5. Sastry R, Mondal S (2013) Geophysical characterization of the Salna Sinking Zone, Garhwal Himalaya, India. Surv Geophys 34:89–119

6. Göktürkler G, Balkaya C, Erhan Z (2008) Geophysical investigation of a landslide: the Altındağlandslide site, İzmir (western Turkey). J Appl Geophys 65:84–96
7. Ogunsuyi O, Schmitt DR, Derek M, Morgan J, Froese C (2011) Geophysical Study of the Peace River Landslide. AAPG Search and Discovery Article #90173 CSPG/CSEG/CWLS GeoConvention 2011, Calgary, Alberta, Canada, 9–11 May 2011
8. Sastry R, Mondal K, Pachauri A (2012) 2D electrical resistivity tomography of a landslide in Garhwal Himalaya. In: 6th international conference & exposition on petroleum geophysics "Kolkata 2006"
9. Jomard H, Lebourg T, Guglielmi Y, Tric E (2010) Electrical imaging of sliding geometry and fluids associated with a deep seated landslide (La Clapière, France). Earth Surf Process Landf 35:588–599
10. Hibert C, Grandjean G, Bitri A, Travelletti J, Malet J-P (2012) Characterizing landslides through geophysical data fusion: example of the La Valette landslide (France). Eng Geol 128:23–29
11. Migoń P, Pánek T, Malik I, Hrádecký J, Owczarek P, Šilhán K (2010) Complex landslide terrain in the Kamienne Mountains, Middle Sudetes, SW Poland. Geomorphology 124:200–214
12. Tric E, Lebourg T, Jomard H, Le Cossec J (2010) Study of large-scale deformation induced by gravity on the La Clapière landslide (Saint-Etienne de Tinée, France) using numerical and geophysical approaches. J Appl Geophys 70:206–215
13. Jessica B, Giuseppe C, Giocoli A, Luongo R, Perrone A, Lapenna V, Piscitelli S (2016) Electrical resistivity tomography surveys for the geoelectric characterization of the Montaguto landslide (southern Italy). Nat Hazards Earth Syst Sci Discuss. https://doi.org/10.5194/nhess-2016-28
14. Akpan E, Ilori O, Essien U (2015) Geophysical investigation of ObotEkpo Landslide site, Cross River State, Nigeria. J Afr Earth Sci 109:154–167. https://doi.org/10.1016/j.jafrearsci.2015.05.015
15. Prodán TH, Prácser E, Balogh J, Kis É, Klébesz R (2015) Geoelectric tomography carried out on a mass movement site of Kulcs settlement (in Hungary). J Geogr Nat Disast 5:149. https://doi.org/10.4172/2167-0587.1000149
16. Shan C, Bastani M, Malehmir A, Persson L, Engdahl M (2014) Integrated 2D modeling and interpretation of geophysical and geotechnical data to delineate quick clays at a landslide site in southwest Sweden. Geophysics 79(4):EN61–EN75. https://doi.org/10.1190/geo2013-0201
17. Ardi N, Iryanti M (2016) Sliding zone identification of landslide area using resistivity method in Cijambe, Subang West Java. AIP Conf Proc 1708:070001. https://doi.org/10.1063/1.4941174
18. Fikosi I, Chitea F (2015) Landslides footprints on geoelectrical data—insights from two case studies. GEO2015—ESG Project Section—20 Nov 2015, Bucharest
19. Ogusnsuyi O (2010) Geophysical characterization of Peace River landslide. M.S. thesis. University of Alberta, Canada
20. Socco LV, Jongmans D, Boiero D, Stocco S, Maraschini M, Tokeschi K, Hantz D (2010) Geophysical investigation of the Sandalp rock avalanche deposits. J Appl Geophys 70:277–291
21. Oppikofer T, Jaboyedoff M, Pedrazzini A, Derron M-H, Blikra LH (2011) Detailed DEM analysis of a rockslide scar to characterize the basal sliding surface of active rockslides. J Geophys Res 116:F02016
22. Fressard O, Maquaire Y, Thiery R, Davidson C Lissak (2016) Multi-method characterisation of an active landslide: Case study in the Pays d'Auge plateau Göktürkler, G., (Normandy, France). Geomorphology 270:22–39. https://doi.org/10.1016/j.geomorph.2016.07.001
23. Chambers J, Meldrum P, Gunn D, Wilkinson P, Kuras O, Weller A, Ogilvy R (2009) hydrogeophysical monitoring of landslide processes using automated time-lapse electrical resistivity tomography (ALERT). Near surface 2009—15th European Meeting of Environmental and Engineering Geophysics, Dublin, Ireland
24. Zhang X (2016) Numerical simulation study on the detection of weak structural plane of rock slope by using 3D electrical resistivity tomography. Engineering 8:438–444. https://doi.org/10.4236/eng.2016.87041

25. Di Maio R, Piegari E (2011) Water storage mapping of pyroclastic covers through electrical resistivity measurements. J Appl Geophys 75:196–202
26. DiMaio R, Piegari E (2012) A study of the stability analysis of pyroclastic covers based on electrical resistivity measurements. J Geophys Eng 9:191–200
27. Shan W, Hu Z, Jiang H, Guo Y, Wang C (2013) Mechanism of permafrost landslide based on GPS and resistivity surveying. Progress of geo-disaster mitigation technology in Asia, Environmental science and engineering. Springer, Berlin, pp 349–361
28. Lebourg T, Hernandez M, Zerathe S, El Bedoui S, Jomard H, Fresia B (2010) Landslides triggered factors analysed by time lapse electrical survey and multidimensional statistical approach. Eng Geol 114:238–250
29. Wilkinson P, Chambers J, Kuras O, Meldrum P, Gunn D (2011) Long-term time-lapse geoelectrical monitoring. First Break 29:77–84
30. Travelletti J, Sailhac P, Malet J-P, Grandjean G, Ponton J (2012) Hydrological response of weathered clay-shale slopes: water infiltration monitoring with time-lapse electrical resistivity. Hydrol Process 26:2106–2119
31. Giocoli A, Stabile T, Adurno I, Perrone A, Gallipoli M, Gueguen E, Norelli E, Piscitelli S (2015) 2015): Geological and geophysical characterization of the southeastern side of the High Agri Valley (southern Apennines, Italy. Nat Hazards Earth Syst Sci 15:315–323
32. Maniruzzaman S, Ramanujam N, Rasool Q, Biswas S, Prasad P, Ojha C (2017) Application of 2D electrical resistivity tomographic imaging technique to study climate induced landslide and slope stability through the analysis of factor of safety: a case study in Ooty Area, Tamil Nadu, India. World Acad Sci Eng Technol Int J Geotech Geol Eng 4(9)
33. Reci H, Muceku Y, Jata Y (2013) The use of ERT for investigation of Berzhita Landslide, Tirana Area, Albania. Landslide Sci Pract 2. https://doi.org/10.1007/978-3-642-31445-2_15. (Springer-Verlag Berlin Heidelberg 2013)
34. Ogilvy R, Kuras O, Meldrum P, Wilkinson P, Chambers J, Sen M, Pulido Bosch A, Gisbert J, Jorreto S, Frances I, Tsourlos P (2009) Automated monitoring of coastal aquifers with electrical resistivity tomography. Near Surf Geophys 7:367–375
35. Supper R, Ottowitz B, Kim J, Römer A, Baron I, Pfeiler S, Lovisolo M, Gruber S, Vecchiotti F (2014) Geoelectrical monitoring: an innovative method to supplement landslide surveillance and early warning. Near Surf Geophys 12(1):133–150
36. Wilkinson P, Chambers J, Meldrum P, Gunn D, Ogilvy R, Kuras A (2010) Predicting the movements of permanently installed electrodes on an active landslide using time-lapse geoelectrical resistivity data only. Geophys J Int 183:543–556
37. Kim J, Yi M, Park S, Kim J (2009) 4-D inversion of DC resistivity monitoring data acquired over a dynamically changing earth model. J Appl Geophys 68:522–532
38. Hayley K, Pidlisecky A, Bentley LR (2011) Simultaneous time-lapse electrical resistivity inversion. J Appl Geophys 75:401–411
39. Jardani A, Revil A, Dupont JP (2012) Stochastic joint inversion of hydrogeophysical data for salt tracer test monitoring and hydraulic conductivity imaging. Adv Water Resour. https://doi.org/10.1016/j.advwatres.2012.08.005
40. Reynolds JM (2011) An introduction to applied and environmental geophysics, 2nd edn
41. Sass O, Bell R, Glade T (2008) Comparison of GPR, 2D-resistivity and traditional techniques for the subsurface exploration of the Öschingen landslide, Swabian Alb (Germany). Geomorphology 93:89–103
42. Lee C, Yang H, Liu H, Wen L, Wang Z, Chen J (2008) A Study of the hydrogeological environment of the Lishan landslide area using resistivity image profiling and borehole data. Eng Geol 98:115–125
43. Panek T, Hradecky J, Šilhan K (2008) Application of electrical resistivity tomography (ERT) in the study of various types of slope deformations in anisotropic bedrock: case studies from the Flysch Carpathians. Stud Geomorphol Carpatho-Balc 42:57–73
44. Jomard H, Lebourg T, Tric E (2007) Identification of the gravitational boundary inweathered gneiss by geophysical survey: La Clapière landslide (France). J Appl Geophys 62:47–57

45. Chang P, Chen C, Chang S, Wang T-B, Wang C-Y, Hsu S-K (2012) An investigation into debris flow induced by Typhoon Morakot in the Siaolin Area, Southern Taiwan, using electrical resistivity imaging method. Geophys J Int 188:1012–1024
46. Zerathe S, Lebourg T (2012) Evolution stages of large deep-seated landslides at the front of a subalpine meridional chain (Maritime-Alps, France). Geomorphology 138:390–403

Analysis of Evaporative Cooling of Inlet Air Influence on Gas Turbine Cycle Performance

Veenit Kumar and Dhaneshwar Mahto

Abstract This paper studies the influence of evaporative cooling air conditioning for a gas turbine. Evaporative cooling makes sure of the evaporation of water to reduce the inlet air temperature of the compressor. The model used for the evaporative cooler panels is based on the energy balance of ambient air and cooling water. Inlet air cooling increases the power output by taking advantage of the gas turbine's feature of higher mass flow rate, due to compressor inlet air temperature decays. As the specific volume of air is directly proportional to the temperature, the increases in the air density result in a higher air mass flow rate, once the volumetric rate is constant. The effects of the parameter (design and operation condition) on the power output, compression work, specific fuel consumption, and thermal efficiency are evaluated and compared with simple gas turbine cycle. Simple gas turbine cycle when precooled its specific work output and efficiency increases as precooling reduces the load on the compressor. But this effect becomes more pronounced when different configurations are used, i.e., when it is required to enhance the specific work output of the cycle reheating is added to the simple cycle and when the exhaust gas of the turbine is used to increase the temperature of the compressed air, i.e., regeneration, it will increase the efficiency of the precooled cycle compared to simple and reheated cycle. Again the effect of intercooling is to increase the specific work output further. Thus, the best possible configuration through which gas turbine performance can be enhanced is intercooling and reheating combined with regeneration.

Keywords Mass flow rate · Specific volume · Specific work output · Efficiency

---

V. Kumar (✉)
Department of Mechanical Engineering, Cambridge Institute of Technology, Ranchi, India
e-mail: veenit.kits@gmail.com

D. Mahto
Department of Mechanical Engineering, Birla Institute of Technology, Mesra, Ranchi 835215, Jharkhand, India

J. Chattopadhyay et al. (eds.), *Renewable Energy and its Innovative Technologies*,
https://doi.org/10.1007/978-981-13-2116-0_17

1 Introduction

The main factors affecting the performance of gas turbines are pressure ratio and turbine working temperature. The difficulty of obtaining a sufficiently high-pressure ratio with an adequate efficiency was not resolved until the application of aerodynamics science to the problem. It is now possible to make advanced engines using pressure ratios up to 35:1, efficiencies of 85–90% and turbine inlet temperature exceeding 1650 K with the development of aerodynamics science, and metallurgy. The maximum fuel/air ratio that may be used is governed by the working temperature of the highly stressed turbine blades. This depends upon the creep strength of the materials used in the construction of the blades. The most used system is the mechanical vapor compression system. However, evaporative cooling as a precooler can be used as an economic alternative for conventional systems. The natural phenomena for the operation of evaporative cooling are induction using water and air as the working fluids. With the passage of airflow utilizing evaporation of water, the air temperature is decreased. Evaporative cooling avoids air dehumidification, a typical discomfort resent in conventional systems. The additional attractiveness of evaporative cooling is its low energy consumption, easy maintenance, and operation.

2 Research Methodology

The density of combustion air and hence the power output are increased by combining evaporative coolers with gas turbines [1]. Gas turbine performance is critically limited by the predominating ambient temperature. It occurs because the ambient temperature and power output are inversely related [2]. In compression and absorption chiller systems, the influence of air cooling intake on the gas turbine performance was studied and compared by two different cooling systems, i.e., evaporative and cooling coil [3]. In the present study, the effect of evaporative cooling is theoretically studied for simple, reheated, regenerated, and intercooled cycle. In the present study, the effect of evaporative cooling is theoretically studied for simple, reheated, regenerated, and intercooled cycle by combining either of the configurations. The evaporative cooler is a wetted rigid media in which water is distributed through the header and where air passes through the wet porous surface part of water is evaporated absorbing the sensible heat from the air, and increasing the relative humidity. The dry bulb temperature is decreased but the wet bulb temperature is not affected. The incoming air and the water in the evaporative cooler may be considered as an isolated system because no heat is added to or removed from the system, and the process of exchanging the sensible heat of the air for latent heat of evaporation from the water is adiabatic.

Evaporative coolers used in the gas turbine are the wetted rigid media type evaporative cooler, and the evaporating medium is a saturated porous pad. The precooler is installed to reduce the air temperature by (adiabatic) humidifying the air. The precooler consists of a control panel, self-supporting stainless housing, the pads, droplet separators, a set of piping, temperature sensors, hygrometer, cabling which is interconnected to pumps, and skid and stainless steel housing.

The net power output obtained from the gas turbine is given by

$$W_{\text{net}} = W_T - W_c$$

where $W_T = \dot{m}t\, C_{\text{pg}}(T_3 - T_4)$,

$\dot{W}c = \dot{m}a\, C_{\text{pa}}(T_2 - T_1)$, C_{pa} is the specific heat of dry air at constant pressure, taken as 1.005 kJ/kg-k, and $\dot{m}t = \dot{m}f + \dot{m}a$

$$\dot{Q} = \dot{m}a\, C_{\text{pg}}\text{avg}(T_3 - T_2),$$

where C_{pg} avg is the flue gas specific heat which is a function of temperature [4], but it is taken as 1.14 kJ/kg-k for calculation.

The cycle efficiency of the plant is given by

$$\eta = \frac{C_{\text{pg}}\,(T_3 - T_4) - C_{\text{pa}}\,(T_2 - T_1)}{C_{\text{pg}}(T_3 - T_2)}$$

Or

$$\eta = 1 - 1/(r)^{\gamma - 1/\gamma}$$

3 Principle of Operation

The precooler is installed to reduce the air temperature by (adiabatic) humidifying the air. This increases the density of air and thus the mass flow of air through the gas turbine which in turn helps the turbine to increase its output over a turbine operates at ambient air condition. The general arrangement of the precooler for the turbine has the following order: Water sump (Bottom section)-Medium sump-Top sump-Roof section-Bottom section-Bracing-Pad supports. The location of precooler is after filtration unit and before intake (Fig. 1), minimum saturation efficiency: 85%, temperature difference: 1.5–2.5 °C between DB and WB, shut off at a temperature below 15 °C (outside), and blow down: 1 time maximum of evaporation rate (Fig. 2).

Cooling efficiency is defined as the rate between the real decrease of the DBT and the maximum theoretical decrease that the DBT could have for cooling efficiency 100% and the outlet was saturated. Here, summer condition (max. temperature and min. RH have been assumed from data).

Cooling efficiency (CE) is given by

$$\text{Cooling Efficiency} = \frac{\text{Tamb} - \text{Tout}}{\text{Tamb} - \text{Twetbulb}}$$

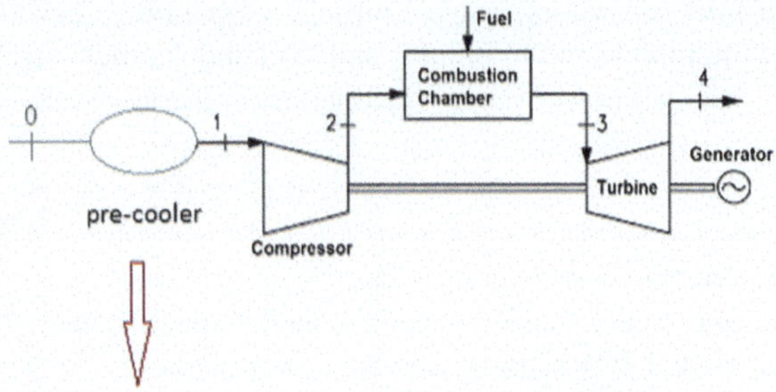

Fig. 1 Precooled gas turbine cycle

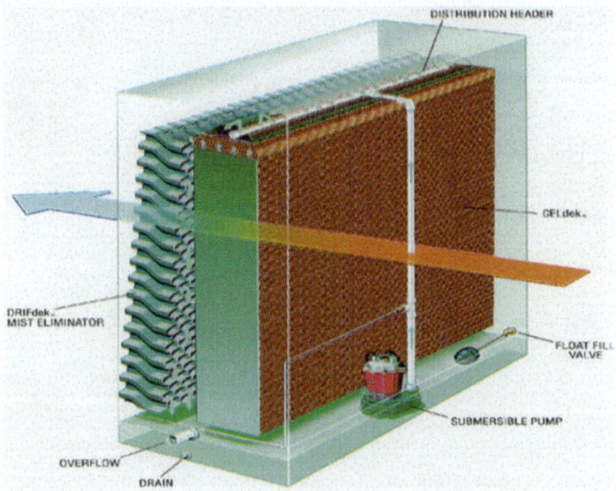

Fig. 2 Evaporative cooler

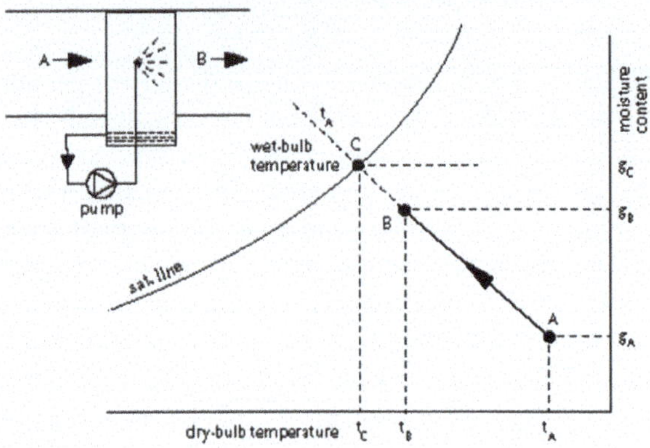

Fig. 3 Psychometric chart for adiabatic humidification of compressor inlet air

T_{amb} = Ambient temperature and $T_{wetbulb}$ = wet bulb temperature or temperature behind coolers. In this present case, the value of $T_{wetbulb}$ = 29.72 °C arrived from psychometric chart assuming ambient conditions, i.e., T_{amb} = 38.0 °C and RH = 56% and the value of Tout, i.e., air leaving the precooler obtained is 31.03 °C (Fig. 3).

Energy balance equation between ambient air and water spread for adiabatic mixing:

$$m_w(hv_3 - hv_2) = m_a(ha_1 - ha_3) + w_1(m_a)(hv_1 - hv_3),$$

and

$$P_{\text{sat}_1} = 610.78 * \text{e}^{(T/T=238.8)} * 17.2694$$

This equation is being taken from reference paper [5], Tetens O.: Ubereinige meterologische Begriffe Zeitschriftfur.

where

hv_3	Enthalpy of water at the outlet,
ha_1	Enthalpy of air at the inlet,
ha_3	Enthalpy of air at the outlet,
$Pv_1 = \phi_1 P_{\text{sat}1}$, where, ϕ_1	Relative humidity of air before precooling,
$(hv_1 - hv_3)$	enthalpy change of air-borne water vapor,

Feed pump work is given by $W_p = m_w \vartheta_{\text{cond}}(P_{\text{in}} - P_{\text{out}})$.

4 Result and Discussions

Independent variable:

Compressor pressure ratio (r)

Dependent variables:

(1) Ambient air temperature (T_1).
(2) Compressor outlet temperature (T_2).
(3) Turbine exhausts temperature (T_4).
(4) Heat supplied (Q).
(5) Specific work output and efficiency.

4.1 Effect of Precooling on Specific Work Output and Efficiency of the Simple Cycle

Here, specific power output is taken as $W_{\dot{n}\text{et}}/C_{\text{pa}}\,T1$ to make it a dimensionless form and the graph is plotted between $W_{\dot{n}\text{et}}/C_{\text{pa}}\,T1$ as a function of pressure ratio. The specific work output for the precooled cycle is increased as, for an engine with constant speed rotation, the volumetric rate is approximately constant. And as a reduction in ambient, air temperature reduces compressor load, which in turn increases $\text{W}_{\dot{n}\text{et}}$ of

Description	Adopted values	Adopted values
Cycle	Simple gas turbine cycle	Precooled gas turbine cycle
Ambient air temperature	311 K	304.03 (cooled by 7 K by evaporative cooling)
Turbine inlet temperature	1350 K	1350 K
Isentropic efficiency of compressor	0.89	0.89
Isentropic efficiency of turbine	0.89	0.89
Combustion chamber pressure loss	2%	2%
Fuel, LHV	Natural gas, 48.235 MJ/kg	Natural gas, 48.235 MJ/kg

Graph 1 Effect on specific work output with increasing pressure ratio of precooled simple cycle

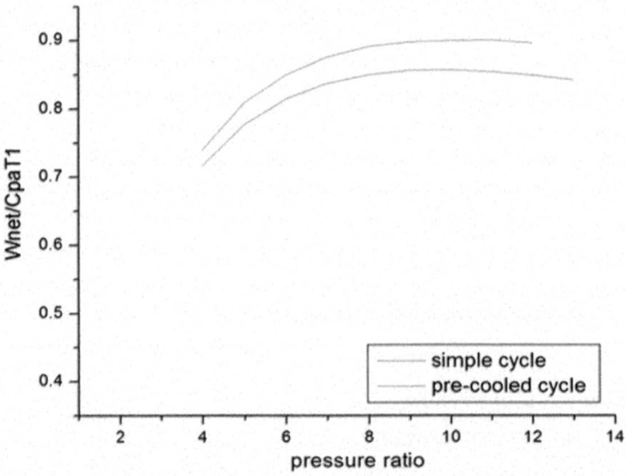

the system. For precooled cycle, the curve has slightly greater values compared to simple cycle as the numerator of the equation is slightly increased for reduction in the value of $T1$ using evaporative cooling as compared to simple cycle, The effect of ambient temperature has been reported by several authors and more recently by Ashley et al. that for every 1 K rise in ambient temperature above ISO condition the gas turbine loses 0.1% in terms of thermal efficiency and 1.47 MW of its Gross (useful) power output (Graph 1).

Graph 2 shows a slight increment in efficiency as the specific power output of the precooled cycle is increased for the same quantity of fuel supplied, as the efficiency mainly depends on the pressure ratio, TIT (turbine inlet temperature), and the nature of gas used.

Graph 2 Effect on efficiency with increasing pressure ratio of precooled simple cycle

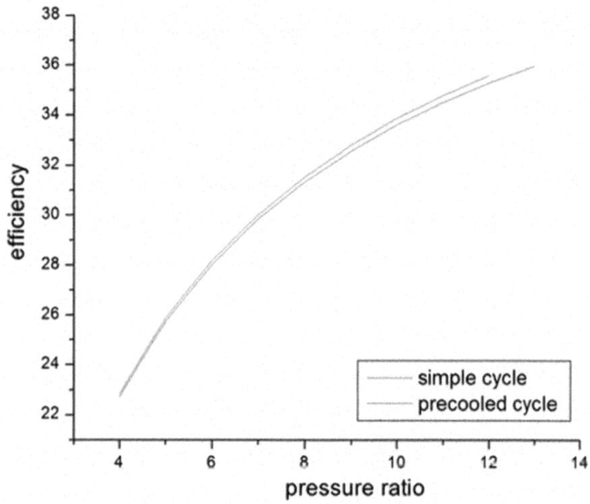

Graph 3 Effect on specific work output with increasing pressure ratio of precooled intercooled cycle

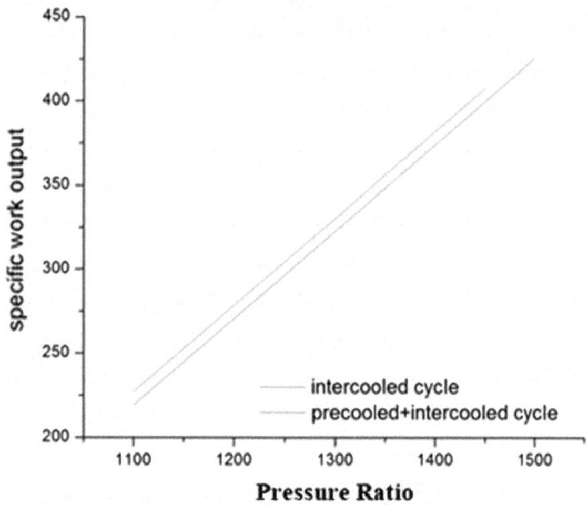

4.2 *Effect of Precooling on Specific Work Output and Efficiency of Intercooled Compression*

It is clear from Graph 3 that the specific work output increases in both cases with increasing pressure ratio. Compression in high-pressure compressor increases net–power output of the cycle and efficiency of the plant actually decreases with increasing pressure ratio as the work supplied to the compressor is saved by inter-cooling but the heat supplied is increased which ultimately decreases the overall efficiency of the plant as shown in Graph 4.

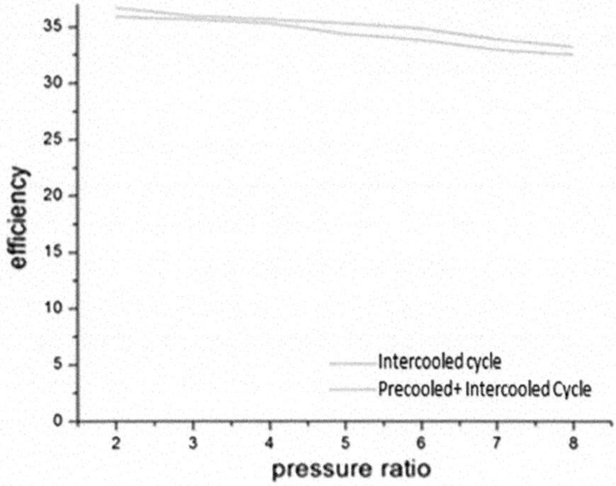

Graph 4 Effect on efficiency with increasing pressure ratio of precooled intercooled cycle

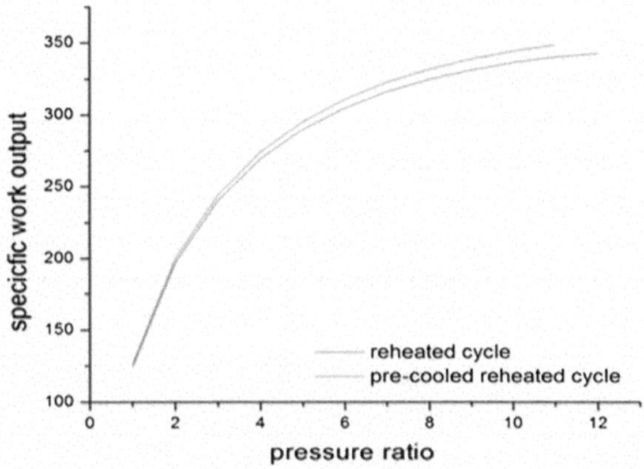

Graph 5 Effect on specific work output with increasing pressure ratio of precooled reheated cycle

4.3 *Effect of Precooling on Specific Work Output and Efficiency of the Reheated Cycle*

Differentiation of the specific work output shows the optimum point in the expansion at which to reheat is when the pressure ratios for H.P. and L.P. turbines are equal. For that optimum value, the reheated cycle will have maximum value for specific work output (Graph 5).

Efficiency has greater values for precooled reheated cycle but actually, the efficiency of gas turbine decreases with reheating, as the two-stage expansion is being counterbalanced by an increase in the quantity of heat supplied which ultimately decrease the efficiency of the plant (Graph 6).

So, there is no benefit in using reheating without regeneration.

Graph 6 Effect on efficiency with increasing pressure ratio of precooled reheated cycle

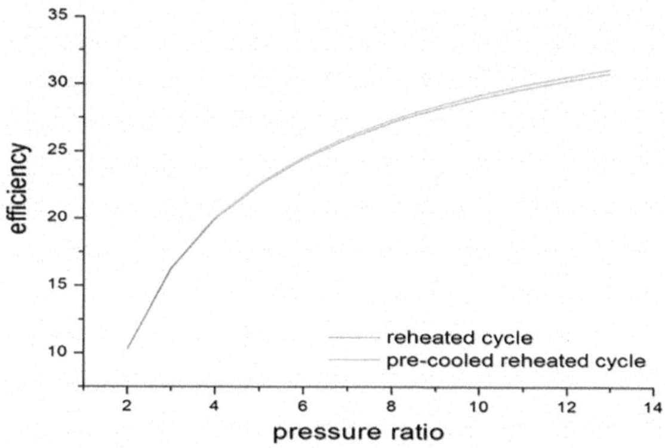

Graph 7 Effect on specific work output with increasing pressure ratio of precooled regenerated cycle

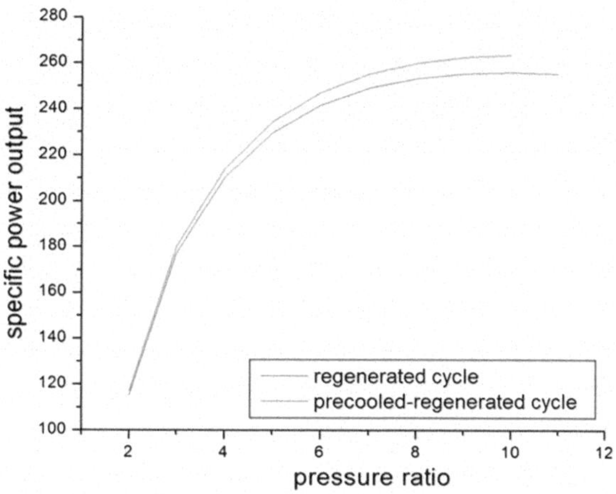

4.4 *Effect of Precooling on Specific Work Output and Efficiency of the Regenerated Cycle*

With increasing pressure ratio, both curves saturate but precooled regenerated cycle has higher values at each point corresponding to the regenerated cycle as the load on the compressor is reduced due to precooling (Graph 7).

Graph 8 shows that the curves fall with an increase in pressure ratio until a value corresponding to $r^{\gamma-1/\gamma} = \sqrt{t}$ is reached, this is the pressure ratio for which the specific work output curve reaches a maximum value, the curve first increases to a max. value called optimum value and thereby decreases, as with increasing pressure ratio the effect of regeneration is to cool the incoming air.

The regenerated cycle attains a maximum value at pressure ratio 6, for the same value of pressure ratio precooled regenerator cycle has higher values.

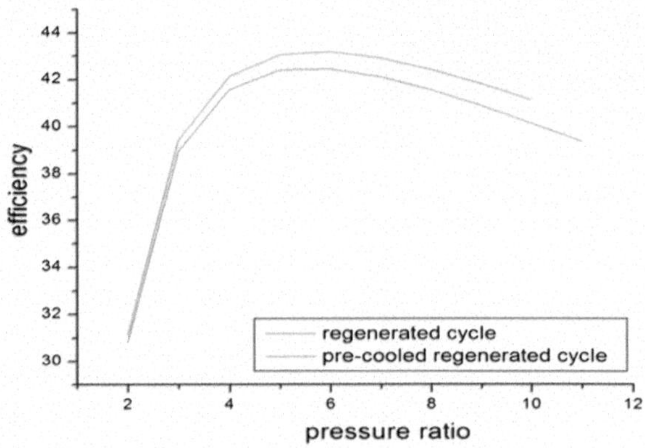

Graph 8 Effect on efficiency with increasing pressure ratio of precooled regenerated cycle

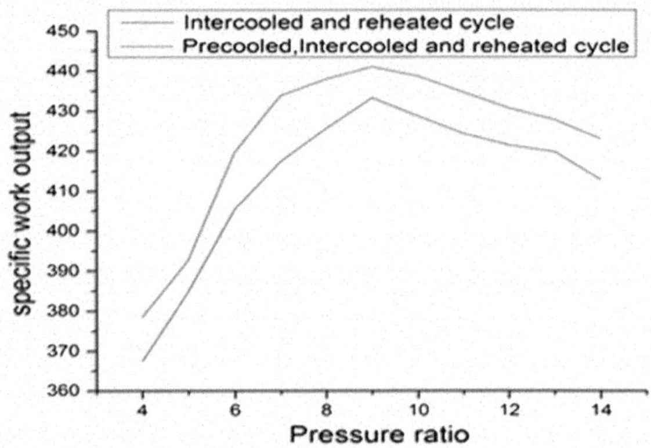

Graph 9 Effect on specific work output with increasing pressure ratio of precooled intercooled compression and reheated cycle

4.5 Effect of Precooling on Intercooled Compression and Reheated Cycle

The specific work output for intercooled and reheated cycle increases to maximum compared with other configurations with the increase in pressure ratio up to optimum point and thereby decreases due to load on compressor is increasing with increase in pressure ratio and when it is precooled the specific work output always has greater values compared to non-precooled cycle and as discussed earlier that intercooling increases the amount of work to be added to the compressor and reheating increases the amount of fuel to be burnt which ultimately leads to lower in the efficiency of the plant (Graphs 9 and 10).

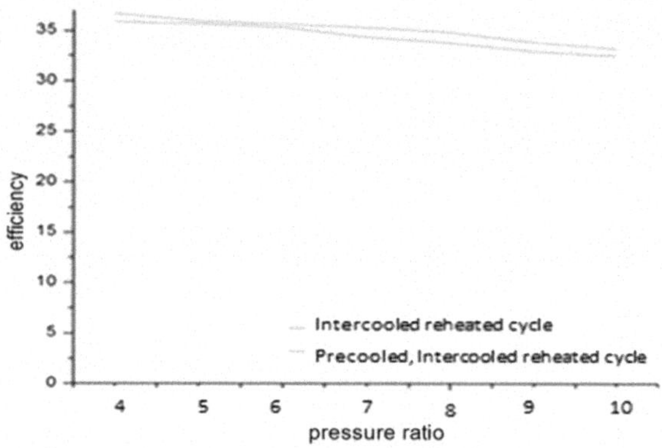

Graph 10 Effect on efficiency with increasing pressure ratio of precooled intercooled compression and reheated cycle

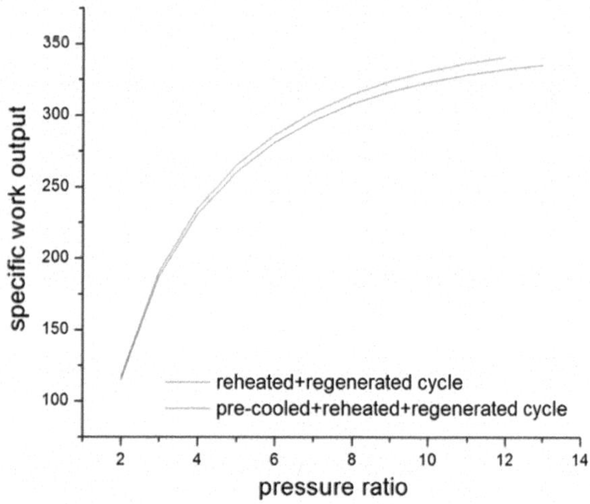

Graph 11 Effect on specific work output with increasing pressure ratio of precooled reheated and regenerated cycle

4.6 Effect of Precooling on Specific Work Output and Efficiency of Reheated and Regenerated Cycle

Graph 11 shows increase in specific work output for precooled reheated and regenerated cycle compared to reheated and regenerated cycle, which further increases with increase in pressure ratio and thermal efficiency first increases with increase in pressure ratio reaches a maximum value and thereafter decreases, the optimum pressure in precooled reheating and regeneration case shifts towards lower pressure ratio and it decreases thereafter due to effect of regeneration. With the increase in pressure ratio, the effect of regeneration is to decrease the temperature of the compressed air (Graph 12).

Graph 12 Effect on efficiency with increasing pressure ratio of precooled reheated and regenerated cycle

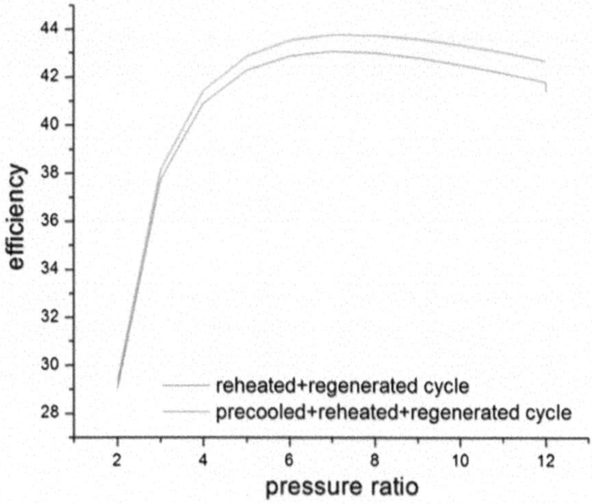

4.7 *Effect of Precooling on Specific Work Output and Efficiency with Increasing Pressure Ratio for Precooled Intercooled and Reheated with the Regenerated Cycle*

Graphs 13 and 14 show the effect of precooling on specific work output and efficiency with the increasing pressure ratio, the specific work output clearly increases as compared to other configurations, and efficiency of the cycle first increases till it acquires an optimum value and then decreases. It is clear from Graph 13 that due to the addition of regeneration, intercooling, and reheating, the peak value of graph shifts towards the lower pressure ratio compared to a simple cycle.

Graph 14 shows the variation of efficiency with increasing pressure ratio, for precooled + intercooled reheated + regenerated cycle, efficiency has a maximum value of 50–51% whereas for those non-precooled cycle, it has a maximum value of 47–48%.

It is evident from Graph 14 that efficiency increases by 1–2% for every 5 °C decrease in ambient air temperature, so in all the industrial applications, a combination of intercooling, reheating and regeneration cycles being used and precooling when combined with this cycle it does help the cause.

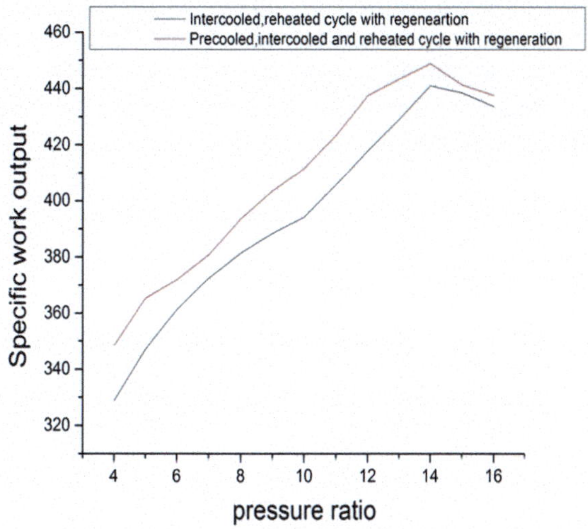

Graph 13 Effect on specific work output with increasing pressure ratio of precooled intercooled compression and reheated cycle with regeneration

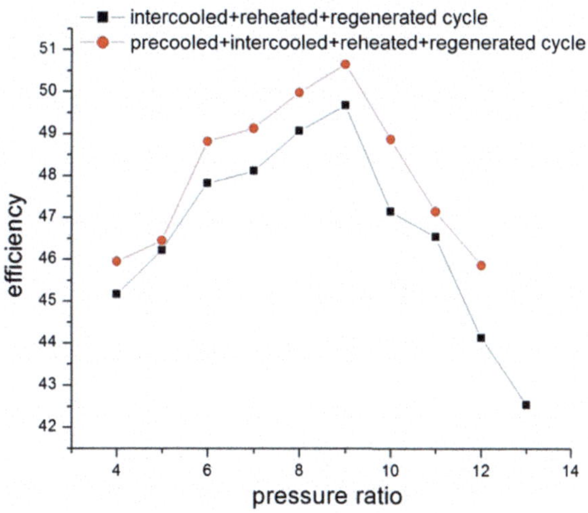

Graph 14 Effect on efficiency with increasing pressure ratio of precooled intercooled compression and reheated cycle with regeneration

5 Conclusion

In this parametric analysis compressor, inlet air is being cooled by evaporative cooling. The different configuration of gas turbine performance has been investigated and compared. Simple gas turbine cycle when precooled, its specific work output and efficiency increase as precooling reduces the load on the compressor. When it is required to enhance the specific work output of the cycle reheating is added to the simple cycle and when the exhaust gas of the turbine is used to increase the temperature of compressed air, i.e., regenerated it will increase the efficiency of precooled cycle compared to simple and reheated cycle. But this effect becomes more pronounced when different configurations are combined. Thus, the best possible configuration through which gas turbine's performance can be enhanced is intercooling and reheating along with regeneration and when it is precooled, it further increases the efficiency of the plant up to 50–51%, 51.02% maximum in this present case concluded from Graphs 13 and 14.

References

1. ASHRAE (2004) Thermal guidelines for data processing environments. American Society of Heating, Refrigeration, and Air-Conditioning Engineers, Atlanta, GA
2. Nasser AEM, El-Kalay MA (1991) A heat-recovery cooling system to conserve energy in gas-turbine power stations in the Arabian Gulf. Appl Energy 38(2):133–142
3. Jaber QM, Jaber JO, Khawaldah MA (2007) Assessment of power augmentation from gas turbine power plants using different inlet air cooling systems. Jordan J Mech Ind Eng 1(1):7–15
4. Alhazmy MM, Najjar YSH (2004) Augmentation of gas turbine performance using air coolers. Appl Therm Eng 24(2–3):415–429
5. Tetens O (1930) Uber einige meteorologische Begriffe. Zeitschrift fur Geophysik 6:297

Biodiesel Extracted from Waste Vegetable Oil as an Alternative Fuel for Diesel Engine: Performance Evaluation of Kirlosker 5 kW Engine

Shiv Kumar Ray and Om Prakash

Abstract These days, we are facing challenges like energy depletion, environmental issue, and rising costs of petroleum, which leads to research for new and renewable energy resources. Among all renewable energy, biodiesel has several advantages as an alternative fuel in terms of availability, performance, and emission characteristics. In this paper, biodiesel is derived from waste vegetable oil and its performance evaluation has been carried on single cylinder diesel engine, make Kirlosker TV1 at 1500 RPM. The diesel engine is performed successfully on all the biodiesel blends without any engine modifications. The highest SFC for biodiesel was 10% more than the diesel. The maximum variation in BTE of diesel has been seen at 1.5 kW loads and was 19.95% as compared to 60% biodiesel blend. At peak load, the air–fuel ratio demand for diesel was 8.83% more than the 100% biodiesel. Up to 20% biodiesel blend, the diesel engine performance was almost similar to diesel without any lag but after that the performance slightly reduces. Hence, biodiesel from waste vegetable oil may be considered as a low-cost renewable substitute for diesel fuel.

Keywords Biodiesel · Diesel engine performance · Waste vegetable oil
Waste frying oil · Waste cooking oil

Nomenclature

v	Volume of fuel
s	Specific gravity
t	Time
sf	Fuel consumption rate

S. K. Ray (✉) · O. Prakash (✉)
Department of Mechanical Engineering, National Institute of Technology Patna, Patna 800005, Bihar, India
e-mail: shivroy2k5@gmail.com

O. Prakash
e-mail: om.prakash@nitp.ac.in

J. Chattopadhyay et al. (eds.), *Renewable Energy and its Innovative Technologies*,
https://doi.org/10.1007/978-981-13-2116-0_18

sfc Specific fuel consumption
bp Brake power output
η Thermal efficiency

1 Introduction

These days, we are facing many challenges like environmental issue and rising costs of petroleum, and also the world's energy demand has enormously increased which leads to the depletion of conventional energy resources [1, 2].

The conventional fuels are the main cause of the air pollution. Therefore, it is the cause of concern which initiates research worldwide for new and renewable energy for sustainable fuel. Biodiesel derived from waste vegetable oil is being explored as a promising alternative to the conventional fuels [3]. Researchers have reported that particulate matter, unburned hydrocarbons, carbon monoxide and sulfur levels are significantly less by using biodiesel, but nitrogen oxide levels get increased with biodiesel [4].

Biodiesel provides lubricating properties that can reduce engine wear and extend engine life [5].

Transesterification of WCO depends on the quality of the feedstock oil [6]. Among all, methanol has been reported to give the best result in terms of biodiesel yield [7]. The biodiesel yields were inversely linked to the acid value and viscosity of the feedstocks [8]. A biodiesel yield of 88–90% can be obtained with methanol and oil ratios of 7:1–8:1 and 0.75 the percentage of KOH [9]. Production of biodiesel by supercritical methanol can obtain a yield of 94% [10]. Waste frying oil requires a heating temperature of 135 °C to bring down its viscosity to that of diesel at 30 °C [11]. Diesel waste cooking oil blends fuel can be directly used in the engine without modification. [12]. Ethanol and ethyl hexyl nitrate can be used as additives in the diesel–biodiesel blends [13]. Engine performance and emissions characteristics of the WCO biodiesel blends were little poorer than diesel, NOx emissions were slightly higher while HC was lower [14]. It is recommended to use cooking oil biodiesel up to 20% with diesel without any engine modifications [15]. The properties of biodiesel and petroleum diesel such as viscosity, specific gravity, flash point, and calorific value, from the reviewed study are given in Table 1 [16].

2 Experimental Setup

The biodiesel derived from waste vegetable oil (WVBD) by transesterification process was blended with petroleum diesel in different volumetric proportions (5, 10, 20, 40, 60, and 80%) and referred as WVBD. For the performance evaluation, a single cylinder, four-stroke direct injection diesel engine, make Kirlosker TV1, 5.2 kW was selected, due to its wide use in agriculture. The engine specifications are given in the

Table 1 Property ranges of fuels used in the reviewed studies [16]

S. No.	Properties	Biodiesel	Diesel
1	Density (15 °C) (kg/m^3)	870–895	810–860
2	Viscosity (40 °C) (cSt)	3.5–5.5	2–3.5
3	Cetane number	45–65	40–55
4	Cold filter plugging point (°C)	−5 to 10	−25 to 0
5	Cloud point (°C)	−5 to 10	−20 to 0
6	Pour point (°C)	−15 to 10	−35 to 0
7	Lower heating value (MJ/kg)	36.5–38	42.5–44
8	Water content (mg/kg)	0–500	–
9	Acid number (mg KOH/g)	0–0.60	–
10	Ester content (% w/w)	>96	–
11	Glycerin content (% w/w)	0–0.25	–
12	Sulfur content (mg/kg)	–	15–500

Table 2 Engine specifications

S. No.	Make	Kirlosker
1	Type	TV1
2	Bore × stroke	87.5 mm × 110 mm
3	Cubic capacity	0.661 L
4	Power	7 BHP (5.2 kW)
5	Rated speed	1500 RPM
6	Compression ratio	17.5:1
7	Injection system	Direct

Table 2. The schematic diagram of the experimental setup can be seen in Figs. 1 and 2. An alternator was coupled with the engine with variable load panel in order to measure the brake power output. The specifications of alternator were given in Table 3. To measure the brake power, engine shaft was coupled to an electrical dynamometer equipped with load panel. Two separate fuel tanks were installed to supply the Diesel and Biodiesel at different proportion. Two burettes each of 50 mL fixed volume was used to measure the fuel consumption rate and for the measurement of airflow rate supplied to the engine an orifice plate consisted with the U-tube manometer has been installed. An air surge tank was installed to minimize the fluctuation of air supply. All experiments were carried out at a constant speed of 1500 RPM.

Fig. 1 Kirlosker TV1 agriculture engine

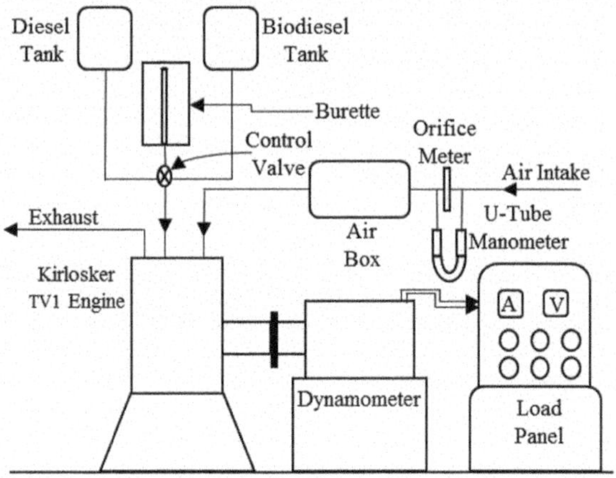

Fig. 2 Schematic diagram of the experimental setup

3 Methodology

For the measurement of fuel consumption rate, specific fuel consumption and brake thermal efficiency, test ware carried out on 50 mL of each fuel at different load and for each test time taken for consumption of fuel by the engine was measured with the help of stopwatch. The mass flow rate of air suction was measured by U-tube manometer fitted with the surge tank orifice.

Table 3 Alternator specifications

S. No.	Make	Kirloskar
1	Model	KBE-105 M
2	Output	5 KVA
3	Volts	230
4	Phase	1
5	AMPS	21.7
6	RPM	1500
7	FREQ	50 Hz

3.1 Fuel Consumption Rate (Fc)

It is a measure of fuel consumption at specific load with respect to time. Equation (1) is used to calculate the fuel consumption rate of the engine for each fuel at different load.

$$\mathrm{fc} = \frac{v \times s \times 3600}{t \times 1000} (\mathrm{kg/h}) \tag{1}$$

where t is the time taken for $v = 50$ mL of fuel consumption, s is the specific gravity of the fuel and fc is the fuel consumption rate.

3.2 Specific Fuel Consumption (SFC)

It is defined as the amount of fuel consumed per hour per unit of brake power output. It is used to compare the performance of different engines. Equation (2) is used to calculate the specific fuel consumption of the engine for each fuel at different load.

$$\mathrm{sfc} = \frac{\mathrm{fc}}{\mathrm{bp}} (\mathrm{kg/kW/h}) \tag{2}$$

where bp is the brake power in kW and fc is the fuel consumption rate in kg/h.

3.3 Brake Thermal Efficiency (η)

It is a measure of how engine converts the heat energy into mechanical energy. Equation (3) is used to calculate the brake thermal efficiency of the engine for each fuel at different load.

$$\eta_{\text{th}} = \frac{\text{bp} \times 3600}{fc \times \text{CV} \times 1000} \times 100 \quad (3)$$

where CV is the calorific value of the fuel kJ/kg and η is the thermal efficiency.

3.4 *Air–Fuel Ratio*

It is the ratio of mass flow rate of fuel to the mass flow rate of air. It was calculated with the help of U-tube manometer filled with water and connected with the orifice of the surge tank on the suction side. It can be seen in Figs. 1 and 2.

$$v_{\text{a}} = C_{\text{d}} A_{\text{o}} \sqrt{\left(2g \Delta h \frac{\rho_{\text{w}}}{\rho_{\text{a}}} \right)} (\text{m}^3/\text{s}) \quad (4)$$

From Eq. (4), the rate of flow of air per unit time is calculated. Where v_a is volume flow rate of air per unit time in m^3/s, C_{d} is the coefficient of discharge, A_o area of orifice, Δh is the monometer difference, ρ_{w} is the density of water, and ρ_{a} is the density of air.

$$m_{\text{a}} = \rho_{\text{a}} \times v_{\text{a}} (\text{kg/s}) \quad (5)$$

From Eqs. (4) and (5), the mass flow rate of air can be obtained.

$$\text{Air fuel ratio} = \frac{m_{\text{f}}(\text{kg/s})}{m_{\text{a}}(\text{kg/s})} \quad (6)$$

From Eqs. (1), (5), and (6), the air–fuel ratio can be obtained.

4 Results and Discussion

4.1 *Specific Fuel Consumption*

The result for the specific fuel consumption of different fuels obtained from the test and calculated from Eq. (2) is given in Table 5. Figure 3 shows the variation of brake specific fuel consumption (SFC) of the engine for diesel and biodiesel with brake power output. The specific fuel consumption decreases with the increase of load and increases with biodiesel percentage. The specific fuel consumption of biodiesel blends was almost same up to 20% biodiesel blend and after that, SFC increases with biodiesel percent. The highest SFC for biodiesel was 10% more than the diesel. This may be due to fuel density, viscosity, and heating value of the fuels.

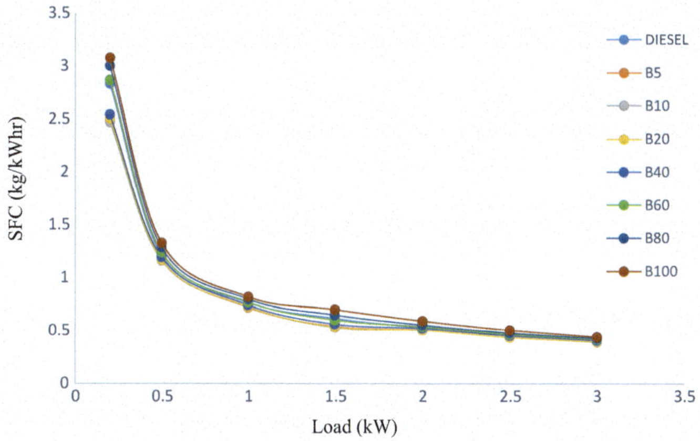

Fig. 3 Specific fuel consumption at different load for different fuels

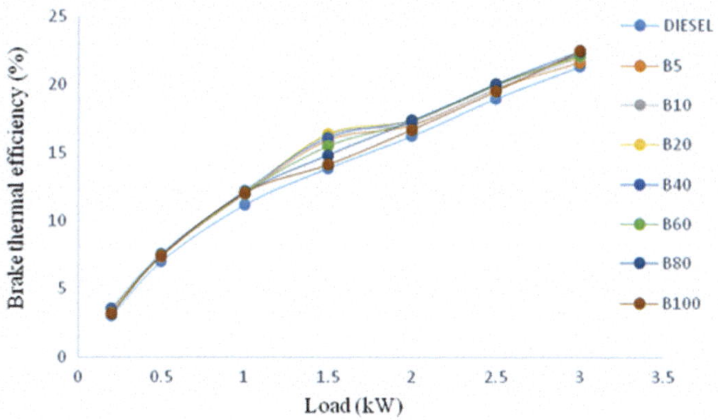

Fig. 4 Brake thermal efficiency at different load for different fuels

4.2 Brake Thermal Efficiency

The result for the brake thermal efficiency of different fuels obtained from the test and calculated from Eq. (3) is given in Table 6. The variation of brake thermal efficiency (BTE) for different blends is given in Fig. 4. It increased with an increase in load for all biodiesel blend and diesel. At pick load, the brake thermal efficiency was almost equal for all biodiesel blends and diesel. This was due to a reduction in heat loss and increase in power with an increase in percent load. The maximum variation in BTE of diesel has been seen at 1.5 kW load and was 19.95% as compared to 60% biodiesel blend.

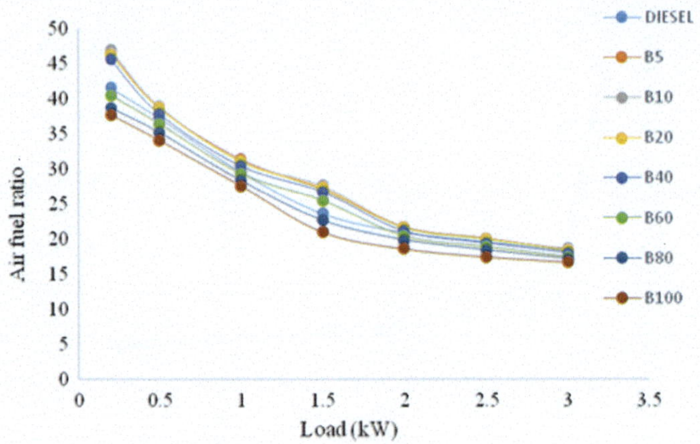

Fig. 5 Air–fuel ratio at different loads for different fuels

4.3 Air–Fuel Ratio

The result for the air–fuel ratio for different fuel obtained from the test and calculated from Eq. (6) is given in Table 7. It can be seen in Fig. 5, at lower load, a higher air–fuel ratio is needed for diesel and biodiesel blends, but as the load increases, the air–fuel ratio decreases. It is because, at low power output, the combustion takes place at lean mixture, but at higher power output, a rich mixture is needed to fulfill the load demand. For all load, the air–fuel ratio was almost same as diesel up to 20% biodiesel blend, and beyond that, it decreases for all blends. At peak load, the air–fuel ratio demand for diesel was 0.83% more than the 100% biodiesel.

5 Conclusions

The diesel engine used in for the experiment has performed successfully with all the biodiesel blends without any engine modifications. Biodiesel was derived from waste vegetable oil and its blends with diesel improved the engine performance when used as a fuel for diesel engine. The specific fuel consumption of biodiesel blends was almost equal up to 20% biodiesel blend and after that, SFC increases with biodiesel percent. The highest SFC for biodiesel was 10% more than the diesel. At peak load, the brake thermal efficiency was almost equal for all biodiesel blends and diesel. The maximum variation in BTE of diesel has been seen at 1.5 kW load and was 19.95% as compared to 60% biodiesel blend. For all loads, the air–fuel ratio was the same as diesel up to 20% and at peak load, the air–fuel ratio demand for diesel was 8.83% more than the 100% biodiesel. Thus, it has been concluded that up to 20% biodiesel blend, the diesel engine performance was almost similar to diesel without any lag but after that, the performance slightly reduces. Hence, biodiesel from waste vegetable oil may be considered as a low-cost renewable substitute for diesel fuel. The use of biodiesel instead of fossil fuels can also help in reducing the environmental pollutions.

Acknowledgements I want to thank teaching and nonteaching staff of Mechanical Engineering Department, National Institute of Technology, Patna, for supporting this work.

Appendix

See Tables 4, 5, 6, and 7.

Table 4 Fuel consumption rate

Load	Diesel	B5	B10	B20	B40	B60	B80	B100
0.2	0.566	0.496	0.494	0.501	0.508	0.573	0.600	0.615
0.5	0.608	0.584	0.581	0.585	0.597	0.622	0.644	0.665
1	0.767	0.721	0.725	0.728	0.743	0.771	0.800	0.820
1.5	0.927	0.814	0.799	0.807	0.845	0.904	0.975	1.051
2	1.057	1.016	1.016	1.014	1.043	1.080	1.112	1.187
2.5	1.128	1.097	1.101	1.098	1.136	1.168	1.200	1.266
3	1.206	1.196	1.184	1.197	1.223	1.264	1.287	1.322

Table 5 Specific fuel consumption

Load	Diesel	B5	B10	B20	B40	B60	B80	B100
0.2	2.832	2.481	2.470	2.506	2.542	2.867	3.001	3.076
0.5	1.217	1.168	1.162	1.169	1.194	1.243	1.288	1.329
1	0.767	0.721	0.725	0.728	0.743	0.771	0.800	0.820
1.5	0.618	0.543	0.533	0.538	0.563	0.603	0.650	0.700
2	0.529	0.508	0.508	0.507	0.522	0.540	0.556	0.593
2.5	0.451	0.439	0.440	0.439	0.454	0.467	0.480	0.506
3	0.402	0.399	0.395	0.399	0.408	0.421	0.429	0.441

Table 6 Thermal efficiency

Load	Diesel	B5	B10	B20	B40	B60	B80	B100
0.2	3.027	3.480	3.521	3.520	3.572	3.260	3.206	3.220
0.5	7.045	7.389	7.485	7.544	7.602	7.518	7.468	7.454
1	11.181	11.981	11.989	12.119	12.220	12.116	12.034	12.076
1.5	13.869	15.899	16.321	16.397	16.119	15.501	14.807	14.144
2	16.214	16.987	17.118	17.392	17.407	17.308	17.301	16.697
2.5	18.996	19.680	19.751	20.081	19.982	20.013	20.050	19.567
3	21.331	21.648	22.038	22.107	22.273	22.176	22.433	22.474

Table 7 Air–fuel ratio

Load	Diesel	B5	B10	B20	B40	B60	B80	B100
0	48.092	48.180	48.762	47.616	47.184	45.956	40.133	39.049
0.2	41.550	46.788	47.004	46.329	45.614	40.491	38.682	37.739
0.5	37.249	38.775	39.009	38.754	37.900	36.456	35.177	34.101
1	29.559	31.437	31.243	31.129	30.462	29.374	28.340	27.622
1.5	23.609	27.109	27.636	27.368	26.756	25.504	22.659	21.023
2	20.890	21.722	21.739	21.772	21.146	20.450	19.857	18.613
2.5	19.579	20.133	20.067	20.110	19.420	18.917	18.410	17.450
3	18.321	18.455	18.658	18.449	18.039	17.468	17.164	16.702

References

1. Angeliki NM (2014) On energy consumption and GDP studies; a meta-analysis of the last two decades. Renew Sustain Energy Rev 29:31–36
2. Vladimir VK, Alexey GT (2010) World energy and climate in the twenty-first century in the context of historical trends: clear constraints to the future growth. J Globalization Stud 20:27–40
3. Agrawal AK, Das LM (2001) Bio-diesel development and characterization for use as a fuel in compression ignition engines. Trans ASME 123:440–447
4. Alessandro BD, Bidini G, Zampilli M, Laranci P, Bartocci P, Fantozzi F (2016) Straight and waste vegetable oil in engines: Review and experimental measurement of emissions, fuel consumption and injector fouling on a turbocharged commercial engine. Fuel 182:198–209
5. Demirbas A (2003) Bio-diesel fuels from vegetable oils via catalytic and non-catalytic supercritical alcohol transesterifications and other methods: a survey. Energy Convers Manag 44:2093–2109
6. Demirbas A (2009) Progress and recent trends in bio-diesel fuels. Energy Convers Manag 50:14–34
7. Banerjee A, Chakraborty R (2009) Parametric sensitivity in the transesterification of waste cooking oil for bio-diesel production: a review. Resour Conserv Recycl 53:490–497
8. Predojevic ZJ (2008) The production of bio-diesel from waste frying oils: a comparison of different purification steps. Fuel 87:3522–3528
9. Phan AN, Tan M (2008) Bio-diesel production from waste cooking oils. Fuel 87:3490–3496
10. Ghoreishi SM, Moein P (2013) Bio-diesel synthesis from waste vegetable oil via transesterification reaction in supercritical methanol. J Supercrit Fluids 76:24–31
11. Pugazhvadivua M, Jeyachandranb K (2005) Investigations on the performance and exhaust emissions of a diesel engine using preheated waste frying oil as fuel. Renew Energy 30:2189–2202

12. Patel MJ, Patel TM, Rathod GR (2015) Performance analysis of engine C.I. using diesel and waste cooking oil blend. IOSR J Mech Civil Eng 12(2):27–33
13. Ramakrishna Srinivas K, Balu N, Naik B, KalyaniRadha K (2013) Performance and emission analysis of waste vegetable oil and its blends with diesel and additive. J Eng Res Appl 3(6):473–478
14. Parekh PR, Goswami J (2012) Emission and performance of diesel engine using waste cooking oil bio diesel blends: a review. J Eng Res Stud 3(1):34
15. Gada MS, El-Bazb FK, Kinawyc OSE (2015) Performance of diesel engines burning used cooking oil (UCO) biodiesel. Int J Mech Mechatron Eng 15(3):74–80
16. Antolin G, Tinaut FV, Briceno Y, Castano V, Perez C, Ramirez AI (2002) Optimization of biodiesel production by sunflower oil transesterification. Bioresour Technol 83:111–114

Shiv Kumar Ray is a Ph.D. research scholar in the Department of Mechanical Engineering, National Institute of Technology, Patna, India. His research area of interests is Internal Combustion Engine and Alternative Fuels. He received his M.Tech from NIT, Patna, India in 2013.

Om Prakash is a Professor in the Mechanical Engineering Department, National Institute of Technology, Patna, India. He received his Ph.D. from Indian Institute of Technology, Delhi. His research area of interests is Power Plant, Heat and Mass Transfer, and Thermal Engineering and Energy.

Effects of Wire EDM Machining Variables on Material Removal Rate and Surface Roughness of Al 6061 Alloy

D. Pramanik, A. S. Kuar and D. Bose

Abstract Wire Electrical Discharge Machining (WEDM) is one of the fastest extensive growing processes in nontraditional machining. So broad are its capabilities that it can encompass the production, aerospace/aircraft, nuclear reactor, medical industries, and virtually all areas of conductive material machining. But low material removal rate and high surface roughness in the WEDM process are the most difficult problems in practical application. Analysis of the effect and the optimization of cutting parameters on material removal rate and surface roughness in the WEDM of Al 6061 t6 alloy have been made in the present paper. Experimental investigation based on Response Surface Methodology (RSM) has been carried out to optimize the main process parameters like pulse on time, pulse off time, wire feed rate, and gap voltage. Experimental validation of the proposed models shows that the desired Material Removal Rate (MRR) and surface roughness (R_a) can be achieved by adjustment of cutting parameters.

Keywords WEDM · RSM · Al 6061 alloy · MRR · R_a

1 Introduction

Wire Electrical Discharge Machining (WEDM) is known to be a thermoelectric nontraditional machining process for conductive material where workpiece material is caused to be eroded by a series of discrete sparks between the workpiece and tool

D. Pramanik (✉) · A. S. Kuar (✉)
Department of Production Engineering, Jadavpur University, Kolkata, India
e-mail: debpramanik18@yahoo.com

A. S. Kuar
e-mail: askuar@rediffmail.com

D. Bose (✉)
Department of Mechanical Engineering, National Institute of Technical Teachers' Training and Research, Kolkata, India
e-mail: dipan88@yahoo.com

J. Chattopadhyay et al. (eds.), *Renewable Energy and its Innovative Technologies*,
https://doi.org/10.1007/978-981-13-2116-0_19

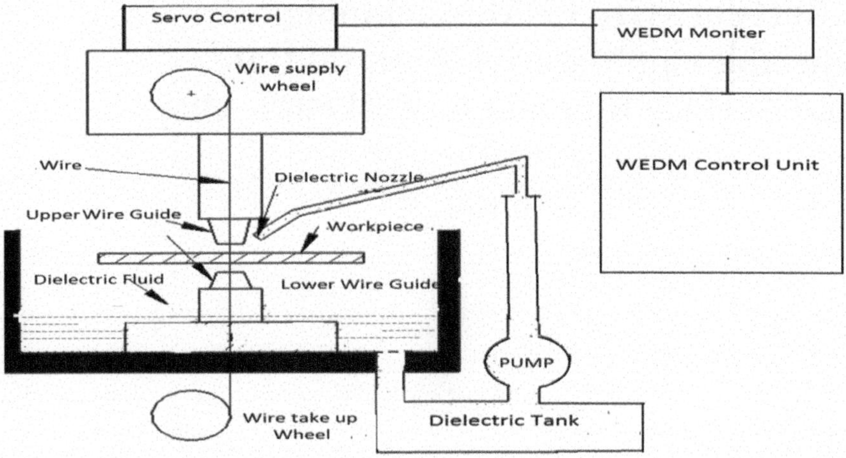

Fig. 1 The schematic diagram of WEDM process

electrode which is plunged in a liquid dielectric medium. Use of a small wire is made as the tool electrode in the Wire EDM process. Normally, deionized water is used as the dielectric medium. The workpiece is placed on the table of the machine and ejection of the dielectric medium is done to the sparking area [1, 2]. A schematic representation of WEDM process is shown in Fig. 1.

Numerical control of the movement of the wire is made to perform the optimum complex shape for the workpiece. Variables, large in number and the stochastic nature of the process hinder achievement of the optimum performance by a highly skilled operator with a state of the art WEDM process. Only the performance of the process and its controllable input parameters may solve the problem effectively with the use of suitable mathematical techniques. For improvement of the desired material removal rate and low surface roughness of Wire Electric Discharge Machining (WEDM) process, cultivation of a number of approaches have been made by the researchers. Tosun et al. [3] have made use of Taguchi's approach to find out and to get optimum effects of dielectric flushing pressure, wire speed, pulse duration, and open-circuit voltage on kerf and material removal rate (MRR) for AISI 4140 steel, from which inference has been made that open-circuit voltage and pulse duration are the highly significant factors for both the MRR and the kerf when less important factors are dielectric flushing pressure and wire speed. Kanlayasin and Boonmung [4] have looked into the influences of four machining variables such as pulse peak current, pulse on time, pulse off time, and wire tension on the surface roughness of wire EDM for DC53 die steel and have evolved a mathematical model by the use of multiple regression method. Material removal rate and surface roughness are very important and plays a critical role in determining the quality and productivity of engineering components. A good quality of surface improves the fatigue strength, corrosion and wears resistance of the product. In this research study, it has been a challenging task to do an experimental investigation on Wire EDM of Al 6061 t6 alloy, a very malleable material having a low melting point. The present study makes a unique approach to investigate the change of main influencing factors of the various machining parameters which are controllable and affect machining achievement on material removal rate and surface roughness. From the equation of developed second-order polynomial, the optimum level of parametric settings is tried to find out to get the desired machining characteristic. The mode of variations of material removal rate and surface roughness is modeled using a statistical approach.

2 Experimentation

Experiments have been conducted on EZEECUT PLUS CNC Wire cut EDM machine manufactured by Ratnaparkhi Electronics India Pvt. Ltd. Aluminum 6061 t6 alloy plate of 5 mm thickness as the workpiece and a brass wire of 0.25 mm diameter as the tool electrode (cathode) are used in this experiment. The central composite rotatable second-order designs guided by response surface methodology constitute the present experiments. A total of 31 experiments have been conducted and is replicated three times. Based on some pilot experiments, the following four parameters, i.e., pulse on time (T_{on}), pulse off time (T_{off}), wire feed rate (WF), and gap voltage (GV) are chosen as machine input. Table 1 shows five levels with four controllable machine parameters which are considered for single-pass cutting operation. Other factors are expected to influence the major of performance and those are kept constant, i.e., product size and shape (rectangular), temperature of dielectric (26 °C), job thickness (5 mm), peak current (3 A), and wire type (0.25 mm diameter of brass). Surface roughness is measured by Mahr Marsurf PS 1 surface texture measuring instrument. A digital micrometer having least count 0.001 mm is used for this purpose. Material Removal Rate (MRR) for the WEDM operation is calculated using formula Eq. (1), which is shown below

$$\text{MRR} = \frac{\text{Total cutting Length (mm)} \times \text{Kerf Width (mm)} \times \text{thickness of Work-piece (mm)}}{\text{Total time taken (min)}} \text{mm}^3/\text{min} \quad (1)$$

Table 1 Controllable process parameters and their limits

Controllable parameters	Units	Levels				
		−2	−1	0	1	2
Wire feed rate (WF)	m/min	35	50	65	80	95
Gap voltage (GV)	Volt	10	20	30	40	50
Pulse on time (T_{on})	μs	12	27	42	57	72
Pulse off time (T_{off})	μs	3	5	7	9	11

The controllable parameters their decided ranges along with the different levels are shown in Table 1.

3 Experimental Results and Discussion

Experimental details for considering the relationship between the controllable process parameters and the various machining criteria has been prepared based on central composite second-order rotatable polynomial design with coded value is shown in Table 2. Table 2 also represents the experimentally obtained results for response 1 and response 2, i.e., Material Removal Rate (MRR) and Surface roughness (R_a), respectively.

3.1 *Mathematical Models for MRR and* R_a

The experimental results consider 4 variables from 31 experiments, the mathematical models for MRR and R_a are developed. Minitab 17 a statistical software is used for analysis of the responses and determining the mathematical models with the best fits are given below as follows Eqs. (2) and (3).

$$\begin{aligned} \mathrm{MRR} = {} & 3.4522 - 0.2130\,\mathrm{WF} - 0.2321\,\mathrm{GV} + 0.0749\,T_{on} - 0.5538\,T_{off} \\ & - 0.0057\,\mathrm{WF} \times \mathrm{WF} - 0.1229\,\mathrm{GV} \times \mathrm{GV} + 0.2248\,T_{on} \times T_{on} \\ & + 0.1924\,T_{off} \times T_{off} - 0.3824\,\mathrm{WF} \times \mathrm{GV} - 0.0753\,\mathrm{WF} \times T_{on} \\ & + 0.1313\,\mathrm{WF} \times T_{off} + 0.2148\,\mathrm{GV} \times T_{on} - 0.1190\,\mathrm{GV} \times T_{off} \\ & - 0.0352\,T_{on} \times T_{off} \end{aligned} \tag{2}$$

$$\begin{aligned} R_a = {} & 4.6960 + 0.1056\,\mathrm{WF} + 0.2119\,\mathrm{GV} + 0.2886\,T_{on} - 0.0333\,T_{off} \\ & + 0.0143\,\mathrm{WF} \times \mathrm{WF} - 0.0839\,\mathrm{GV} \times \mathrm{GV} - 0.1624\,T_{on} \times T_{on} - 0.0235\,T_{off} \times T_{off} \\ & - 0.0049\,\mathrm{WF} \times \mathrm{GV} - 0.0880\,\mathrm{WF} \times T_{on} - 0.2567 \times T_{off} + 0.0033\,\mathrm{GV} \times T_{on} \\ & - 0.2858\,\mathrm{GV} \times T_{off} + 0.3203\,T_{on} \times T_{off} \end{aligned} \tag{3}$$

3.2 *Analysis of Variance and Model Fitment Test*

To establish the mathematical link between the responses and machining parameters of WEDM process, the analysis of variance test has been performed to test the adequacy of the developed model. Design of the ANOVA test module has been made in such a way that it can estimate the sum of square of the response into the contribution due to the second order and lack of fit component, measuring the deviations of the responses from the fitted surface simultaneously with the measure of the experimental error. Tables 3 and 4 represent the ANOVA for MRR and R_a, respectively.

Table 2 DOE and experimental results

Experimental information					Results	
S. No.	Machining parameters (coded value)				Material removal rate (MRR)	Surface roughness (R_a)
	WF	GV	T_{on}	T_{off}		
1	50	20	27	5	4.392	3.496
2	80	27	27	5	4.584	4.428
3	50	40	27	5	4.501	4.562
4	80	40	27	5	3.193	5.465
5	50	20	57	5	4.238	3.665
6	80	20	57	5	4.240	4.248
7	50	40	57	5	5.204	4.631
8	80	40	57	5	3.600	5.153
9	50	20	27	9	3.387	3.924
10	80	20	27	9	4.081	3.838
11	50	40	27	9	2.883	3.740
12	80	40	27	9	2.240	3.580
13	50	20	57	9	3.037	5.279
14	80	20	57	9	3.540	4.775
15	50	40	57	9	3.616	5.159
16	80	40	57	9	2.445	4.739
17	35	30	42	7	3.957	4.606
18	95	30	42	7	3.068	4.989
19	65	10	42	7	3.481	3.978
20	65	50	42	7	2.606	4.832
21	65	30	12	7	4.150	3.513
22	65	30	72	7	4.719	4.668
23	65	30	42	3	5.446	4.692
24	65	30	42	11	3.163	4.601
25	65	30	42	7	3.194	4.700
26	65	30	42	7	3.534	4.609
27	65	30	42	7	3.414	4.752
28	65	30	42	7	3.519	4.702
29	65	30	42	7	3.509	4.688
30	65	30	42	7	3.508	4.721
31	65	30	42	7	3.486	4.698

Table 3 Analysis of variance for Material Removal Rate

Source	DF	Adj. SS	Adj. MS	F-value	p-value
Model	14	16.6393	1.18852	96.80	0.000
Linear	4	9.8770	2.46924	203.32	0.000
WF	1	1.0889	1.08886	106.06	0.000
GV	1	1.2923	1.29233	91.69	0.000
T_{on}	1	0.0134	0.01346	1.15	0.010
T_{off}	1	7.3611	7.36111	614.39	0.000
Square	4	3.0716	0.76790	34.67	0.000
2-way interaction	6	3.6907	0.61511	61.28	0.000
Error	16	0.2534	0.01584		
Lack of fit	10	0.1666	0.01666	1.15	0.451
Pure error	6	0.0869	0.01448		
Total	30	16.8927			

$S = 0.125858$; R-sq = 98.50%; R-sq(adj) = 97.19%; R-sq(pred) = 93.62%

Table 4 Analysis of variance for surface roughness

Source	DF	Adj. SS	Adj. MS	F-value	p-value
Model	11	8.41702	0.76518	221.47	0.000
Linear	4	3.37065	0.84266	243.90	0.000
WF	1	0.26774	0.26774	77.49	0.000
GV	1	1.07734	1.07734	311.83	0.000
T_{on}	1	1.99890	1.99890	578.56	0.000
T_{off}	1	0.02667	0.02667	7.72	0.012
Square	3	0.91954	0.30651	88.72	0.000
2-way interaction	4	4.12683	1.03171	298.62	0.000
Error	19	0.06564	0.00345		
Lack of fit	13	0.05415	0.00417	2.17	0.174
Pure error	6	0.01149	0.00417		
Total	30	8.48267	0.00192		

$S = 0.0587788$; R-sq = 99.23%; R-sq(adj) = 98.78%; R-sq(pred) = 97.78%

From Table 3, it is observed that the associated p-value of less than 0.05 for the model (i.e., $\alpha = 0.05$) shows that the model are statistically significant. Both the developed second-order polynomial regression model for this responses and interactions of parameters are important. The implication of F-value (1.15) and p-value (0.451) of the lack of fit remains that it's significant is not relative to the pure error, as this is desired. From the table, it is observed that statistically; pulse off time (74.52%) is the most dominating factor followed by wire feed rate (11.02%), gap voltage (13.08%), and pulse on time (1.36%).

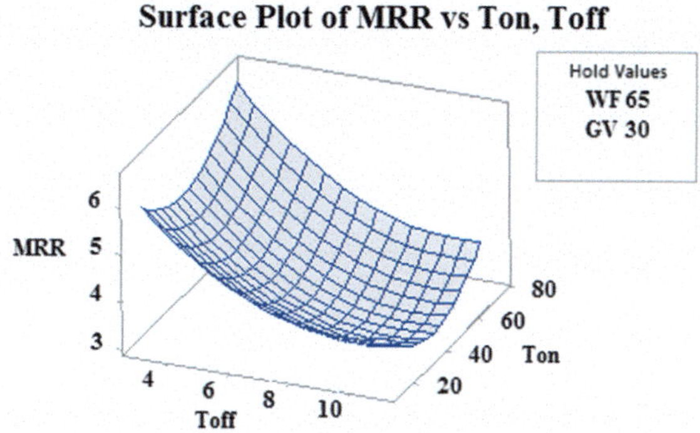

Fig. 2 Surface plot of Material Removal Rate (MRR) versus pulse on time (T_{on}) and pulse off time (T_{off})

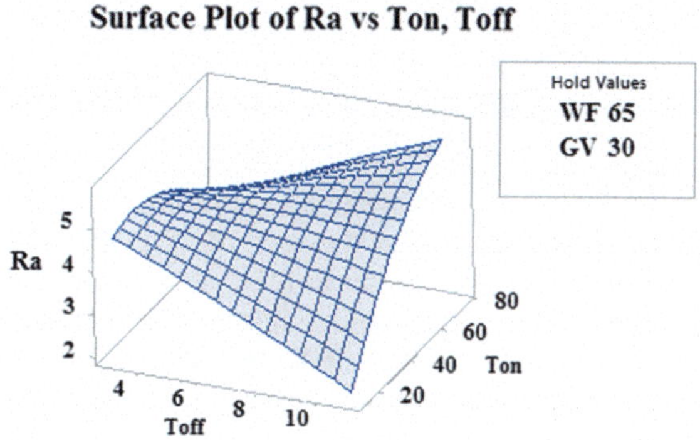

Fig. 3 Surface plot of surface roughness with pulse on time (T_{on}) and pulse off time (T_{off})

From Table 4, it is observed that the associated *p*-value of less than 0.05 for the model (i.e., $\alpha = 0.05$) shows that the model are statistically significant. From the table, it is noted that statistically; pulse on time (59.30%) is the most dominating factor followed by gap voltage (31.96%), wire feed rate (7.94%), and pulse off time (0.79%).

From the surface plot in Fig. 2 shows the combined effect of pulse on time and pulse off time on Material Removal Rate (MRR). The wire feed rate and gap voltage are kept constant at 65 m/min and 30 V, respectively. It is observed that Material Removal Rate is increased with increase of pulse on time and gradually decreases with the implementation of pulse off time. This is because a higher value of pulse on time generates high machining time, which results in high MRR.

From the surface plot in Fig. 3 shows the combined effect of pulse on time and pulse off time on Surface Roughness. It is evident from the Fig. 3 that surface roughness is decrease with decrease of pulse on time and gradually increases with the implementation of pulse off time. At low pulse off time means less machining time as a result surface roughness is quite low. A higher pulse on time generates higher thermal energy on the material surface. For surface roughness, high value of pulse on time, results in a bigger crater on the surface. It shows that pulse on time and pulse off time have the extremely important effect on surface roughness (R_a).

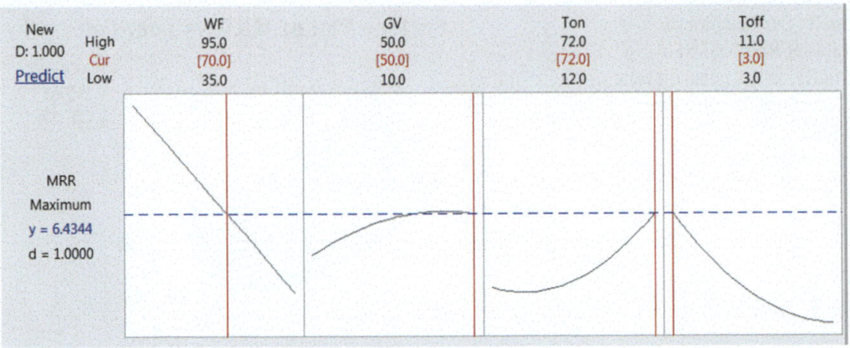

Fig. 4 Optimization result of material removal rate (MRR)

3.3 Optimal Process Parameter of MRR

Figure 4 shows the optimization results for maximum material removal rate based on the developed mathematical model, i.e., Equation (2). In case of linear desirability function (*d*), the value is set as 1, i.e., their working rate covers all parameters. For maximization of the response, similar importance is kept on the lower, target, and the upper bound of the linear desirability function. The parameters settings for maximum material removal rate has been shown to be 6.4344 mm^3/min when wire feed rate 70 m/min, gap voltage 50 V, pulse on time 72 μs, and pulse off time 3 μs.

3.4 Optimal Process Parameter of Surface Roughness (R_a)

Figure 5 shows the optimization results for minimum surface roughness based on the developed mathematical model, i.e., Equation (3). In case of linear desirability function (*d*), the value is set as 1, i.e., their working rate covers all parameters. For maximization of the response similar importance is kept on the lower, target, and the upper bound of the linear desirability function. The parameters settings for minimum surface roughness has been shown to be 2.8568 μm when wire feed rate 40 m/min, gap voltage 45 V, pulse on time 20 μs, and pulse off time 10 μs.

After getting the optimal parameter settings, the next step is to verify the feasibility of the proposed response surface equations [5]. It is observed from the validation experiments that there is a small percentage error between the estimated and the experimental values, which suggest that the developed models can yield nearly exact results within the limits of cutting parameters being used (Table 5).

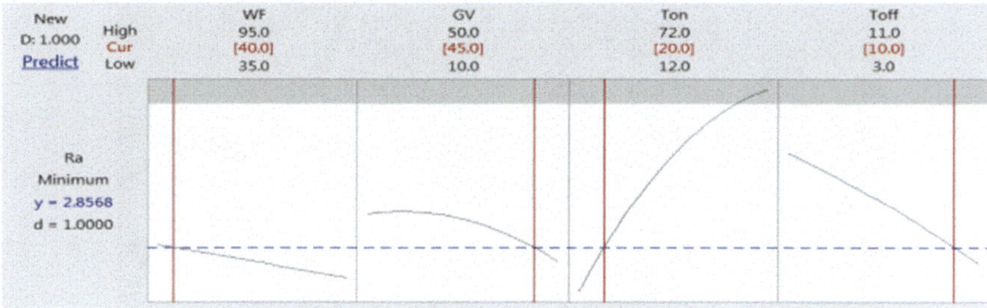

Fig. 5 Optimization result of surface roughness (R_a)

Table 5 Summary of the results of verification experiments

Responses	Actual value	Predicted value	Difference	% Error
Material removal rate	6.4579	6.4344	0.0235	2.35
Surface roughness	2.8924	2.8568	0.0356	3.56

4 Microstructure Analysis

Discharge action in WEDM causes the craters on the workpiece surface. Dielectric fluid and electrode materials also affect it. Metallurgical changes are visible in the upper layer of the workpiece as the temperature of the discharges reaches 8000–12,000 °C. A thin recast epitaxial layer is produced as a result of melting of the surface of the workpiece material and quick rehardening of it due to cooling action of the dielectric.

Here also, it has been shown before machining of surface topography in Fig. 6. Only some casting defects are there.

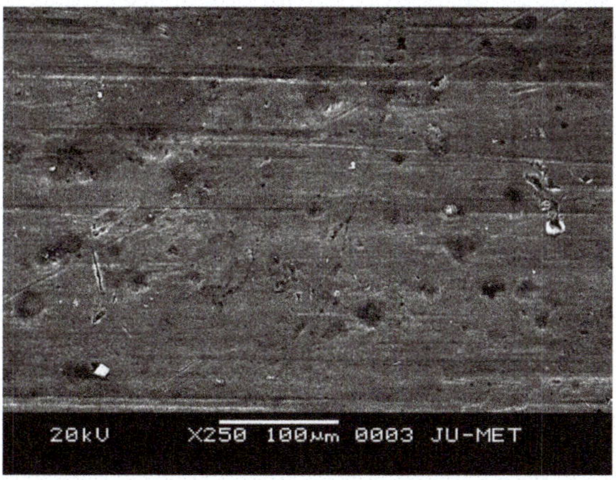

Fig. 6 Scanning electron micrograph of metal surface topography (before machining)

Fig. 7 SEM photograph of the wire EDM cut surface of a sample machined under the following machine settings: wire feed rate = 65 m/min, gap voltage = 30 V, pulse on time = 57 μs, and pulse off time = 9 μs

Investigation of the surface topography of WEDM surfaces and exploration of possible ways to regulate its parameters to perform better surface roughness are the two objectives of the present study. Application of a response surface methodology is made to analyze surface topography with wire feed rate, gap voltage, pulse on time, and pulse off time as input parameters. Scanning Electron Micrographs (SEMs) showed the complex appearance of the surfaces with shallow craters, spherical bubbles, melt drops, globules of debris, micro cracks, and voids arising out of high heat energy released by discharges and subsequent cooling. The spherical particles appear to be molten metals, which have their random expulsion at the time of discharge and then their solidification and attachment to the surface. The pulse on time and gap voltage have been observed as the most significant parameters affecting the surface properties. When pulse on time was increased, the surface texture of the machined surface is composed of varying size of deep craters rims. For successive electrical discharges, intense heat and local melting or evaporaziation of work material crater rims deep and overlapping—are formed (Fig. 7).

5 Conclusion

In the present research, Response Surface Methodology is used for investigating the effect on the main parameters that affect the machining criteria such as material removal rate and surface roughness. Almost, all the control factors have been considered simultaneously to establish the trends of variations. ANOVA test has been applied successfully to check the adequacy of the machining criteria. The effects of cutting parameters on MRR and Ra in the WEDM of Al 6061 t6 alloy are experimentally investigated for cutting operation. It can be concluded that MRR and R_a

are controlled under certain machining condition wire feed rate, gap voltage, pulse on time (T_{on}), and pulse off time (T_{off}) can be varied as per requirement to achieve maximum MRR and minimum R_a in WEDM. The predicted values of MRR and R_a agree very well with estimated from the experimental results under optimum parameter settings. Achievement of a higher MRR and good surface phenomena at the same time is hardly possible. The most dominant factors for Material Removal Rate (MRR) are pulse on time, pulse off time, wire feed rate. MRR is significantly affected by the higher value of pulse on time and lower value of pulse off time and interaction between wire feed rate and gap voltage. The optimum value of surface roughness has been calculated at lower value of pulse on time and higher value of pulse off time.

References

1. Spedding TA, Wang ZQ (1997) Parametric optimization and surface characterization of wire electrical discharge machining process. Precis Eng 20:5–15
2. Chen Z, Huang Y, Zhang Z, Li H, Ming W, Zhang G (2014) An analysis and optimization of the geometrical inaccuracy in WEDM rough cutting. Int J Adv Manuf Technol 74:917–929
3. Tosun N, Cogunb C, Tosun G (2004) A study on kerf and material removal rate in wire electrical discharge machining based on Taguchi method. J Mater Process Technol 152:316–322
4. Kanlayasiri K, Boonmung S (2007) Effects of wire-EDM machining variables on surface roughness of newly developed DC 53 die steel: design of experiments and regression model. J Mat Process Tech 192:459–464
5. Pramanik D, Kuar AS, Sarkar S, Mitra S, Bose D (2016) Experimental investigation on corner accuracy in WEDM for aluminium alloy, AIMTDR 2016, ISBN: 978-93-86256-27-0

Author Index

J. Chattopadhyay et al. (eds.), *Renewable Energy and its Innovative Technologies*,
https://doi.org/10.1007/978-981-13-2116-0

The manufacturer's authorised representative in the EU is Springer Nature Customer Service Centre GmbH, Europaplatz 3, 69115 Heidelberg, Germany. If you have any concerns regarding our products, please contact ProductSafety@springernature.com

Printed and bound by CPI Group (UK) Ltd, Croydon, CR0 4YY
25/09/2026
02222841-0003